Biodiversity in Locally Managed Lands

Special Issue Editors

Jeffrey Sayer
Chris Margules

MDPI • Basel • Beijing • Wuhan • Barcelona • Belgrade

MDPI

Special Issue Editors

Jeffrey Sayer
James Cook University, Australia
Tanah Air Beta, Indonesia

Chris Margules
James Cook University, Australia
University of Indonesia, Indonesia

Editorial Office
MDPI AG
St. Alban-Anlage 66
Basel, Switzerland

This edition is a reprint of the Special Issue published online in the open access journal *Land* (ISSN 2073-445X) from 2016–2017 (available at: http://www.mdpi.com/journal/land/special_issues/biodiversity_managed).

For citation purposes, cite each article independently as indicated on the article page online and as indicated below:

Author 1; Author 2. Article title. *Journal Name* **Year**, *Article number*, page range.

First Edition 2017

ISBN 978-3-03842-454-3 (Pbk)
ISBN 978-3-03842-455-0 (PDF)

Table of Contents

About the Special Issue Editors

Jeff Sayer is a British ecologist who began his working life researching large mammal ecology in Africa. Research and conservation activities in Africa brought him into daily contact with the realities of the lives of poor rural people trying to make a living in the wildlife rich areas around parks and reserves. He subsequently worked in nature conservation projects in Afghanistan, Myanmar and Thailand where similar conflicts between conservation and poverty alleviation were widespread. He later spent more than 10 years in Indonesia as founding director general of CIFOR (Center for International Forestry Research) where he championed a multi-disciplinary ecosystem approach to sustainable forest management. He has worked at various times for WWF and initiated the Forest Conservation Programme of IUCN. He is now Professor of Conservation and Development at James Cook University in Cairns, Australia, where he is a member of the Center for Tropical Environmental and Sustainability Science and directs a Graduate Programme in Development Practice.

Chris Margules is Adjunct Professor at James cook University in Australia and Research Associate at the Research Center for Climate Change, University of Indonesia. His current interest focuses on integrating conservation and development at the landscape or seascape scale. Previously, he designed and implemented a large-scale experiment on the ecological effects of habitat fragmentation and played a key role in discovering and then implementing the idea of complementarity in systematic conservation planning. He was a research scientist and later research manager at the Commonwealth Scientific and Industrial Research Organisation (CSIRO) for 32 years, where he led programs on landscape management and sustainable development in the tropics. He joined Conservation International in 2006 and later became Senior Vice-President and leader of the Asia Pacific Division. He has worked in number of Asian countries, the Pacific region and southern Africa as well as Australia. He was a fellow of the Wissenschaftskolleg zu Berlin in 1994. He received Order of Australia honours in the General Division (AM) for services to science in 2005.

Preface to "Biodiversity in Locally Managed Lands"

Recent decades have seen a rapid uptake of schemes to hand over forest management to local communities and/or local governments. It has frequently been claimed that local people will be better at managing these forests than companies or more centralized government forest departments. The claims of the advantages of local management may have merit for the broader objectives of forest management but this book questions whether local communities or local governments will be motivated or even have the knowledge and capacity to manage forests for their biodiversity values. The papers suggest that it cannot automatically be assumed that local management will protect all native biodiversity and there are many instances where it clearly will not do this. The book presents evidence of what does and does not work when local people are entrusted with the management of biodiverse forests and concludes that local management may or may not provide the best biodiversity outcomes—everything depends on context. The need to monitor biodiversity in these forests and to adapt management to ensure that biodiversity is protected is highlighted.

Jeff Sayer and Chris Margules

Special Issue Editors

land

MDPI

Article

Biodiversity in Locally Managed Lands

Jeffrey Sayer [1,2,*] and Chris Margules [1,2,3]

1   Center for Tropical Environmental and Sustainability Science, James Cook University, Cairns 4870, Australia;
     chrismargules@gmail.com
2   Tanah Air Beta, Batu Karu, Tabanan, Bali 82152, Indonesia
3   Research Center for Climate Change, University of Indonesia, Depok 16424, West Java, Indonesia
*   Correspondence: jeffrey.sayer@jcu.edu.au

Received: 2 May 2017; Accepted: 11 June 2017; Published: 15 June 2017

Abstract: Decentralizing natural resource management to local people, especially in tropical countries, has become a trend. We review recent evidence for the impacts of decentralization on the biodiversity values of forests and forested landscapes, which encompass most of the biodiversity of the tropics. Few studies document the impact of decentralized management on biodiversity. We conclude that there may be situations where local management is a good option for biodiversity but there are also situations where this is not the case. We advocate increased research to document the impact of local management on biodiversity. We also argue that locally managed forests should be seen as components of landscapes where governance arrangements favor the achievement of a balance between the local livelihood values and the global public goods values of forests.

Keywords: biodiversity; forest landscapes; local management; public goods; livelihoods

---

Human societies constantly make decisions that lead to the loss of biodiversity. The Convention for the Conservation of Biological Diversity in its "ecosystem principles" recognizes that biodiversity conservation is a matter of societal choice. Some societies may choose to forego economic benefits in order to protect rare species, while other societies may choose to maximize growth. Governments can legislate to place severe restrictions on any actions that endanger biological diversity but another democratically elected government, even in the same country, may choose to lessen these restrictions. The Trump regime in the USA is revoking strict laws protecting biodiversity in order to favor job creation. There is widely held acceptance that it will not be possible to conserve all biodiversity in a world populated by 9.5 billion people, most of whom aspire to extravagant levels of material consumption. Societies in effect make choices about how much biodiversity to conserve and how to achieve this. Unfortunately, most of these choices are made on the basis of weak evidence and are driven by emotions and short-term material needs.

Tropical forests are amongst the most species-rich ecosystems on the planet and any management interventions have impacts on biodiversity. When we choose to harvest timber from a forest, we change the species composition of that forest and possibly cause some species to be lost. When a forest is designated as a protected area, we may interrupt natural cycles of fire and ecological succession and have long-term impacts on the species composition of the forest. Clearing a forest for oil palm obviously reduces species diversity. Modern societies take many actions that change the ecological processes of tropical forests and lead to changes in their species composition. The loss of biodiversity that results from all these interventions in tropical forests is widely regarded as a major global environmental challenge [1].

The special issue of *Land* entitled "Biodiversity in Locally Managed Lands" is a response to a recent move to decentralize forest management and conservation to local communities and local governments. The past decade has seen a powerful tendency to pass control of forests and forested landscapes from central government authorities to administrations at more local scales. There has been a rapid expansion of indigenous reserves, community forests, protected landscapes, managed resource areas, and other forms of local management systems that apply at sub-national scales [2,3]. In Indonesia, for example, these might be provinces, districts within provinces, or regencies within districts. We define local involvement as the input to decisions on the use, including conservation, of natural resources by the people who live in and utilize a particular forest or landscape for their livelihoods. These local people come from a variety of different backgrounds and represent a variety of different views on what management outcomes should be. They include men and women, ethnic minorities and majorities, poor people and rich people, people of different religions, and so on. The challenge in managing forests and forested landscapes is to reconcile the heterogeneous array of values that such people hold and find a way forward that most people can agree to [4]. These same issues apply to seascapes, but here we focus on forested landscapes.

The Convention on biological Diversity has adopted the "Aichi Targets" for protected areas, and its member states have set a target of conserving 17% of terrestrial land areas for biodiversity conservation. However, this target is being met not through expansion of national parks and strict nature reserves but by designating many areas where humans practice agriculture and other economic activities as IUCN category V protected areas; protected landscapes and seascapes. Dudley et al. [2] reviewed published information and case studies in an effort to determine the extent to which such areas protect biodiversity. Their evidence is limited and contradictory. Sometimes protected landscapes are apparently better than more strictly protected areas, sometimes they are worse, and sometimes they are just the same. The difficulty in pinning down sound evidence one way or the other reflects a common thread throughout this special issue: that not enough is currently known to say with any certainty that locally managed areas are better or worse at protecting biodiversity [5]. Context is everything.

The impetus for decentralized management is coming largely from civil society organizations whose mandate is to champion the rights of local and indigenous communities [5,6]. We have encountered numerous examples of community organizations who claim that local management provides the best option for conserving biodiversity, but we have found very few scientific studies in the peer-reviewed literature that substantiate these claims. The evidence available is unclear and can be contradictory [2,7]. The objective of this special issue is emphatically not to take a position opposing local forest management. There are many excellent reasons why local people should have major decision-making power over their forests. There are numerous examples of local people who have managed forests and agroforests sustainably for generations but who have been deprived of their resources by corporate land grabs. Local self-determination, especially of the many poor people who live in, and depend upon, tropical forests, should take precedence over rather abstract global biodiversity goals. There are many situations where local management may provide better prospects for biodiversity than any of the likely alternative management options for a forest, but there is little concrete evidence that this is the case.

We simply do not know in most cases what will happen to biodiversity when forests are handed over to local communities or their local governments. The local context will drive outcomes. Governance arrangements, the functions of institutions, powerful local individuals, and the extent to which ethnic minorities are recognized all play significant roles. There are few examples of locally managed forests being subjected to the scientific monitoring that would provide evidence on the fate of biodiversity. Galbraith et al. [8] ask if engaging local communities in project management, a form of citizen science, leads to enduring support for ecological restoration, the control of invasive species, or the protection of native biodiversity in New Zealand. They found that of 50 local groups participating in such projects, none identified strategic milestones or measures of progress towards biodiversity goals. They suggest that improved training, more technical support, and institutional collaboration are all needed to move such local groups towards a more genuine citizen science capability.

We argue that the move to shift control of forests to more local levels must be accompanied by greatly expanded efforts to conduct inventories of biodiversity in these areas and to monitor the impacts of local management strategies on biodiversity. Fujiki et al. [9] describe a method for predicting tree community composition, an indicator of forest intactness that could be used to monitor progress towards the achievement of the Aichi targets. It might also be used to monitor the effects of different management practices at a more local level. Thackway and Freudenberger [10], using examples from Australia, show that the vegetation condition is an emergent property, not only of environmental conditions, but importantly also of markets, technology, history of settlement, and infrastructure development, government policies and programs, and individual and community values, and how all of this is constantly changing over time. They propose a simple graphical report showing drivers of change and trends against a benchmark state, which might be appropriate for monitoring local management practices. It is also a strong argument for strengthening multi-sectoral governance because institutions, powerful individuals, and politics have a profound impact on the fate of biodiversity. If community-managed areas do prove to be significant for biodiversity conservation, then the values that communities provide to broader society through their conservation actions may merit them being rewarded by significant payments for environmental services. If such win–win outcomes are to be realized, then we will need to greatly expand our investments in documenting the biodiversity values of locally managed areas. Changes in biodiversity in locally managed areas will have to be carefully monitored to enable management to be adapted to achieve specific biodiversity outcomes.

In the right circumstances, locally managed lands can clearly make significant contributions to the conservation of biodiversity. Mukul et al. [11] measured the conservation value of four local agroforestry land uses and the forest itself, in and around a protected forest in Bangladesh. They found that agroforestry can complement the protected area. Betel leaf agroforestry protects more biodiversity than pineapple or lemon agroforestry, which in turn are better than shifting agriculture with fallow. Thus, local agroforestry management does contribute to regional biodiversity. However, caution is required. Velho et al. [12] found that, while many native species are present in the locally managed lands adjacent to three protected areas in Northeast India, the larger mammals were generally absent, being restricted to the protected areas. There is clearly a role for locally managed lands in protecting regional biodiversity, but they complement protected areas—they do not substitute for them.

There are numerous examples of communities causing losses to biodiversity. Communities often derive few instrumental benefits from conserving biodiversity and often incur significant costs [13]. Terborgh et al. [3] provide evidence of the loss of keystone species in areas under local management, which leads to cascading impacts on tree diversity. They were not surprised by this, pointing to the fact that people who have lived more or less isolated from world markets have recently become connected to them and now aspire to become modern consumers enjoying the benefits of modern technology, just like the rest of us. In the same vein, we should not be surprised to find that local management priorities are not necessarily aligned with public good outcomes like the protection of biodiversity [6]. Boedhihartono [5] shows that even relatively remote and isolated communities in Indonesia's forests have little incentive to conserve all components of biodiversity. Short-term improvements in livelihoods tend to be prioritized over more abstract long-term biodiversity outcomes. Langston et al. [14] found that forests on the frontier of estate crop plantations in West Kalimantan, Indonesian Borneo, that are still managed locally are coming under increasing pressure. Poverty and aspirations to participate in the cash economy are driving change away from traditional local management practices. Mosaics of different land uses from protection to intensive cultivation and multi-sectoral coordinated governance that such landscapes would require are suggested as possible ways forward.

Advocates of local management should also be more realistic about the long-term sustainability of these systems. Populations are growing, people are more connected to the outside world, and they are better informed about the material benefits that come with economic growth. In a world of 9.5 million people, there will only be few who choose to subsist from hunting and gathering or even by harvesting non-timber products from agroforests. Handing over control of forests to such isolated peoples today may enable them to navigate their transition to a market economy in the future more easily and without losing their main capital asset—their land. However, it seems unlikely that local management alone will provide for the security and prosperity that most of the rural poor have as their preferred long-term objective.

Local involvement in management of natural resources is incontestably desirable and has been under-valued in forest and land conservation strategies in the past. However, local management is not a panacea [15]. Local communities should be central to all decision-making on land allocation, but measures need to be taken to conserve the public goods values of forests—values that accrue to non-local stakeholders. Governance systems that recognize local values must be developed [16]. One of the problems of international initiatives to favor local management is that they often fail to accord with the realities of local governance, which can be complex and, as noted above, multi-sectoral. Hodge et al. [17] contrast short-term project outcomes with longer-term strategies that promote adaptive multi-sectoral governance. Both have their strengths but the second is needed for lasting change because, among other things, it necessarily requires the participation of local communities. Short-term projects are often insufficiently informed by local contexts [18]. Landscape approaches provide one way to ensure that local interests are fully incorporated into decision-making on forest lands [19] and theories of change can ensure that learning and adaptation can enable local interests and global biodiversity goals to be reconciled [4].

The history of forest conservation has been rich with silver-bullet solutions imposed from outside. Many of these solutions have failed because they were not embedded in local realities. Local involvement in forest management and local control of a proportion of all forests is essential if sustainable outcomes that improve local livelihoods and protect biodiversity are to be achieved. Local management has to be an important component of future efforts to conserve forest biodiversity, but it has to be part of a much broader suite of approaches that ensure that the full range of forest values are managed in ways that meet the needs of the broader societies that have legitimate interests in the maintenance of forest values. Local management should not be seen as replacing the need for conventional, strictly protected areas. Instead, locally managed forests should be seen as significant components of multi-functional landscapes that achieve a balance between meeting livelihood needs of local people and safeguarding the public goods values of forests for the benefit of the broader community.

Acknowledgments: The authors were supported by the Centre for Tropical Environmental and Sustainability Science at James Cook University and the NGO, Tana Air Beta. CM acknowledges support from the University of Indonesia.

Author Contributions: Jeffrey Sayer conceived the idea for the paper and the special issue of *Land* that follows. Jeffrey Sayer and Chris Margules wrote the paper and have jointly edited the special issue.

Conflicts of Interest: The authors declare no conflict of interest.

References

1. Sayer, J.; Whitmore, T. Tropical moist forests: Destruction and species extinction. *Biol. Conserv.* **1991**, *55*, 199–213. [CrossRef]
2. Dudley, N.; Phillips, A.; Amend, T.; Brown, J.; Stolton, S. Evidence for biodiversity conservation in protected landscapes. *Land* **2016**, *5*, 38. [CrossRef]
3. Terborgh, J.; Peres, C.A. Do community-managed forests work? A biodiversity perspective. *Land* **2017**, *6*, 22. [CrossRef]
4. Sayer, J.A.; Margules, C.; Boedhihartono, A.K.; Sunderland, T.; Langston, J.D.; Reed, J.; Riggs, R.; Buck, L.E.; Campbell, B.M.; Kusters, K.; et al. Measuring the effectiveness of landscape approaches to conservation and development. *Sustain. Sci.* **2016**, *12*, 1–12. [CrossRef]
5. Boedhihartono, A.K. Can community forests be compatible with biodiversity conservation in Indonesia? *Land* **2017**, *6*, 21. [CrossRef]
6. Sayer, J.; Margules, C.; Boedhihartono, A.K. Will biodiversity be conserved in locally-managed forests? *Land* **2017**, *6*, 6. [CrossRef]

7. Roe, D.; Day, M.; Booker, F.; Zhou, W.; Allebone-Webb, S.; Kümpel, N.; Hill, N.A.; Wright, J.; Rust, N.; Sunderland, T.C.; et al. Are alternative livelihood projects effective at reducing local threats to specified elements of biodiversity and/or improving or maintaining the conservation status of those elements?: A systematic review protocol. *Environ. Evid.* **2014**. [CrossRef]

8. Galbraith, M.; Bollard-Breen, B.; Towns, D.R. The community-conservation conundrum: Is citizen science the answer? *Land* **2016**, *5*, 37. [CrossRef]

9. Fujiki, S.; Aoyagi, R.; Tanaka, A.; Imai, N.; Kusma, A.D.; Kurniawan, Y.; Lee, Y.F.; Sugau, J.B.; Pereira, J.T.; Samejima, H.; et al. Large-scale mapping of tree-community composition as a surrogate of forest degradation in bornean tropical rain forests. *Land* **2016**, *5*, 45. [CrossRef]

10. Thackway, R.; Freudenberger, D. Accounting for the drivers that degrade and restore landscape functions in australia. *Land* **2016**, *5*, 40. [CrossRef]

11. Mukul, S.A.; Saha, N. Conservation benefits of tropical multifunctional land-uses in and around a forest protected area of Bangladesh. *Land* **2017**, *6*, 2. [CrossRef]

12. Velho, N.; Sreekar, R.; Laurance, W.F. Terrestrial species in protected areas and community-managed lands in Arunachal Pradesh, Northeast India. *Land* **2016**, *5*, 35. [CrossRef]

13. Ostrom, E. A general framework for analyzing sustainability of social-ecological systems. *Science* **2009**, *325*, 419–422. [CrossRef]

14. Langston, J.D.; Riggs, R.A.; Sururi, Y.; Sunderland, T.; Munawir, M. Estate crops more attractive than community forests in West Kalimantan, Indonesia. *Land* **2017**, *6*, 12. [CrossRef]

15. Ostrom, E.; Janssen, M.A.; Anderies, J.M. Going beyond panaceas. *Proc. Natl. Acad. Sci. USA* **2007**, *104*, 15176–15178. [CrossRef]

16. Nagendra, H.; Ostrom, E. Polycentric governance of multifunctional forested landscapes. *Int. J. Commons* **2012**, *6*, 104–133. [CrossRef]

17. Hodge, I.; Adams, W.M. Short-term projects versus adaptive governance: Conflicting demands in the management of ecological restoration. *Land* **2016**, *5*, 39. [CrossRef]

18. Sayer, J.A.; Wells, M.P. The pathology of projects. In *Expecting the Unattainable: The Assumptions Behind ICDPs*; McShane, T.O., Newby, S.A., Eds.; Columbia University Press: New York, NY, USA, 2004.

19. Sayer, J.A. Reconciling conservation and development: Are landscapes the answer? *Biotropica* **2009**, *41*, 649–652. [CrossRef]

land

Article

Do Community-Managed Forests Work?
A Biodiversity Perspective

John Terborgh [1,2,]* and Carlos A. Peres [3]

[1]   Nicholas School of the Environment, Duke University, Durham, NC 27708, USA
[2]   Florida Museum of Natural History, University of Florida, Gainesville, FL 32611, USA
[3]   School Environmental Sciences, University of East Anglia, Norwich NR4 7TJ, UK; c.peres@uea.ac.uk
*   Correspondence: manu@duke.edu; Tel.: +1-352-477-5015.

Academic Editors: Jeffrey Sayer and Chris Margules
Received: 3 January 2017; Accepted: 17 March 2017; Published: 27 March 2017

Abstract: Community-managed reserves (CMRs) comprise the fastest-growing category of protected areas throughout the tropics. CMRs represent a compromise between advocates of nature conservation and advocates of human development. We ask whether CMRs succeed in achieving the goals of either. A fixed reserve area can produce only a finite resource supply, whereas human populations exploiting them tend to expand rapidly while adopting high-impact technologies to satisfy rising aspirations. Intentions behind the establishment of CMRs may be admirable, but represent an ideal rarely achieved. People tied to the natural forest subsist on income levels that are among the lowest in the Amazon. Limits of sustainable harvesting are often low and rarely known prior to reserve creation or respected thereafter, and resource exhaustion predictably follows. Unintended consequences typically emerge, such as overhunting of the seed dispersers, pollinators, and other animals that provide services essential to perpetuating the forest. CMRs are a low priority for governments, so mostly operate without enforcement, a laxity that encourages illegal forest conversion. Finally, the pull of markets can alter the "business plan" of a reserve overnight, as inhabitants switch to new activities. The reality is that we live in a hyperdynamic world of accelerating change in which past assumptions must continually be re-evaluated.

Keywords: extractive reserves; communal forests; human-occupied protected areas; Amazonia; indigenous reserves; tropical forest; sustainable-use reserves; hunting; deforestation

Community-managed forests comprise the fastest growing category of protected areas throughout the tropics and now far exceed strictly protected reserves in both numbers and total area (Schmitt, et al., 2009 [1]; Peres 2011 [2]). Yet their role in protecting full complements of tropical biodiversity remains contentious and poorly understood. We provide a conservation biologist perspective on this issue. Both of us have conducted ecological research in Amazonia for several decades, JT in Perú and CAP in Brazil. Our comments are based on long personal experience with native Amazonians, global trends and an understanding of the ecological requirements of biodiversity conservation. Although we shall offer examples from other ecosystems, our emphasis will be on the Amazon where huge tracts of land have been allocated to community management in the form of indigenous reserves and several categories of legally occupied community-managed reserves. For example, a total of 22.0% of the Brazilian Amazon has been allocated to "extractive" and "sustainable development" reserves (de Marques, et al. 2016 [3]), and a further 22.3% is represented by officially designated indigenous territories (RAISG 2015 [4]).

Community-managed reserves in Amazonia were created to sustain the traditional lifestyles and economies of established Amazonians, including both *caboclos* and *ribereños* (people of mostly mixed ethnic background who speak Portuguese or Spanish) and native Americans. Most such people subsist on small slash-and-burn plots supplemented with hunting and fishing. Cash income, if any, is derived from selling forest products, including fish, game (bushmeat) and non-timber products, such as natural rubber, Brazil nuts and several oil seeds. Various restrictions and regulations apply but a constant among them is that there should be no large-scale deforestation.

Amazonian lands designated for community management were intended to be models of sustainable development. When reserves were established, most of the residents lacked capital and modern technology and survived on meager subsistence economies in remote corners of a region the size of the continental US. However, the justifying assumptions of sustainability and perpetuation of traditional lifestyles are being rapidly eroded by the penetration of modernity into the farthest reaches of the basin. In this context, the notion of sustainability has become an oxymoron. A truly sustainable lifestyle, as we view it, would resemble that practiced by native Amerindians prior to Western contact. Human population densities were mostly <1 individual per km^2, technological innovation was glacially slow, and the demographic processes of birth and death were more or less in balance. These fundamental conditions for sustainability have been massively disrupted by government policies designed to lift marginal populations out of poverty and by the rapid expansion of heavily capitalized logging, mining and agriculture into the heart of Amazonia. In other words, all bets are off with respect to the nominal motivation for creating community reserves, that of preserving traditional economies and lifestyles. Conserving biodiversity was always regarded as a secondary benefit of reserve creation and few or no provisions for biodiversity conservation were written into the enabling legislation. In effect, community-managed reserves are free-wheeling entities that operate under few restraints that might help ensure the perpetuation of biodiversity.

At the most basic level, community management and biodiversity conservation are inherently in conflict because the community subsists on the resources of the reserve. It entails exploitation and exploitation implies disturbance and selective removal of preferred resources while others are ignored, a process that progressively enhances the abundance of ignored resources at the expense of preferred ones. The process often leads inexorably to one or both of two alternatives. Either the community expands the area over which it extracts preferred resources and/or it initiates a process of intensification. Intensification can take many forms, such as agroforestry, expansion of forest clearings for agriculture, sowing pasture grasses for livestock, replacing natural forest with tree plantations (e.g., oil palm, rubber, *Eucalyptus*), and aquaculture. None of these forms of intensification enhances native biodiversity and most are detrimental to it.

In the literature on community management, a topic that is almost never mentioned is human demography. Demography is the proverbial elephant in the room. Conservation planning is typically based on current circumstances, not those of even one generation forward when the community is likely to contain twice as many families. It is also based on current lifestyle and access to technology, ignoring the fact that lifestyles inexorably evolve and technology is subject to a process of continuous upgrading and refinement. This brings us to the well-known IPAT equation: human Impact = Population $\times$ Technology $\times$ Consumption, which helps frame how we think about humanity's footprint on the natural environment at different scales (Ehrlich 2014 [5]).

Demographic trajectories of rural and indigenous communities in tropical developing countries often diverge from those in cities. In urban areas, everything has to be bought with hard-earned money. Children are demanding and expensive, and require significant parental investments in education extending over many years, so there are strong incentives to limit family size. But in a rural context, children are mostly regarded as assets because they often contribute labor to family activities instead of going to school. In this context, a local annual population growth rate of ~3.52%, representing a village doubling time of 20 years, could be defined as normal (Joppa, et al. 2009 [6]). How often is a community management project planned on the assumption of a 20-year population doubling time? Not often. And while management plans of communal reserves, if any, may limit or preclude immigration from other areas, they rarely if ever address internal population growth.

In our hyperkinetic world, few people live in such remote locations that they lack access to the latest technology. Highly desired items such as guns, chainsaws, boat motors and smartphones thus find their way into the heart of any African rainforest and the most distant tributaries of the Amazon. People who a generation ago wore grass skirts, paddled dugout canoes and hunted with bows and arrows today wear Western dress, drive outboard motors and hunt with shotguns. Governments are eager to provide public amenities, including roads, primary schools, health clinics, electricity, and more recently cell phone service and the internet. Under the relentless pressure of modernization, nothing stays the same, and material aspirations understandably escalate.

Historically, indigenous communities were autonomous and largely self-sufficient, but exposure to the market awakens desires, which have irreversible consequences for human behavior. A person who walks barefoot, for example, will envy the person who has a pair of shoes. But a pair of shoes can't be made from forest products, it has to be bought, so the individual is confronted with a need for cash. In a remote community, there is unlikely to be readily available employment, so money can only be earned by commercializing products obtained from the communal holdings.

And thus begins a spiraling economy of exploitation that is often inherently unsustainable and that has no logical or practical end point other than the exhaustion of desirable resources. Rules and regulations are rarely enforced because governments are reluctant to curtail the earning capacity of people who are poor, and technical assistance from government agencies or conservation NGOs to ensure the co-management and sustainable harvest of resources is rarely available.

Back to the equation. None of the components of the $I = PAT$ equation are constant or even plausibly constant under any reasonable scenario. Family planning services are often unavailable to rural people, so birth rates remain high. Exposure to television and social media arouses desires and aspirations that motivate the rapid adoption of new technologies and patterns of consumption, making a mockery of any claim of sustainable development.

In areas under community-based management of forest lands, bushmeat, other nontimber forest products and timber are the near-universal sources of cash revenue. In Indonesia, for example, there are specialty products like rattan, durian and gaharu, but rattan and gaharu have become scarce and durians are increasingly grown in plantations (Soehartono, et al. [7]). Few people can any longer make a living extracting them from the natural environment. There is an inevitable progression here, driven by human striving.

Products derived from communal reserves are of two basic types. Either they can be propagated *en masse* in gardens, plantations, or animal enclosures, or they possess properties inimical to mass propagation. Although there have been scores of attempts to domesticate wild animals that are favored in the bushmeat trade, few of these have been successful. More generally, there is a widespread notion that bushmeat can be sustainably harvested, but this is a simplistic view that overlooks the wide spectrum of life-history traits in tropical forest vertebrates, and the extreme vulnerability of species possessing low reproductive potential. For example, one of the most highly favored species of bushmeat across the Amazon is the spider monkey (*Ateles* spp.). The natural interbirth interval in this species is 30 months, longer than in humans (Symington 1987 [8]). Spider monkeys thus cannot withstand even light hunting pressure, and should at best be described as a "bad" game species. Only after exhaustion of the bushmeat resource can protein-starved villagers be persuaded to engage in animal husbandry. However, small-scale animal husbandry is a poor competitor against commercial livestock production, which tends to supply the protein requirements of market-integrated communities much more efficiently.

Prime, slow-growing timber species—such as mahogany and rosewood—are everywhere extracted to exhaustion, even if the harvest in community-managed forests is subject to fairly strict controls under government-enforced reduced-impact logging guidelines (Richardson and Peres 2016 [9]). High-value hardwoods can be difficult to propagate and do not attain commercial size for a century or more, so are beyond the pale for an impoverished villager who has to feed his family today. Low-fecundity game species and slow-growing prime timber are the first to go, and can be effectively defined as non-renewable resources. They should not be considered anything else. To do so is to promote an illusion, a feel-good fiction that does a disservice to the whole concept of sustainable development.

Perhaps the most spectacular example of a sustainable resource extracted from nature was that of natural rubber. It generated vast wealth (and gruesome human exploitation) for a brief period of a few decades until it ended in a crash. The crash followed when plantations in Malaysia—based on rubber tree (*Hevea* spp.) seeds smuggled from Brazil—came on line. Plantations proved to be so much more productive and efficient than collecting rubber from the Amazonian forest that they caused the price to collapse and government subsidies to be discontinued, and with it the whole economy of gathering rubber from the natural forest.

After the murder of the renowned environmental activist Chico Mendes, there were well-intentioned efforts to revive the extraction of natural rubber from the Amazon, including the establishment of several large forested extractive reserves in the Brazilian states of Acre, Amazonas and Amapá. These efforts enjoyed modest success because the price of natural rubber gathered from the forest was heavily subsidized by the Brazilian government. As a member of GATT, the global trade organization, Brazil was eventually required to abandon the subsidy and rubber tapping in the Amazon came to an end. Many of the *seringueiros*, as they are called, were left without a means of support and had to migrate to the towns where most of the Amazon population now lives (Parry et al. 2010 [10]), and the rural exodus still continues today.

At present, the only major non-timber product of global commerce to be extracted from natural tropical forests is Brazil nuts. Brazil-nut trees (*Bertholletia excelsa*) grow to prodigious size, take many decades to mature and have special pollination requirements, so have not been widely propagated in plantations. But grafting and other modern silvicultural techniques are being applied to the challenge of growing smaller, fast-growing Brazil-nut trees in plantations, so it is very likely that the harvest of Brazil nuts from the wild will, as in the case of so many other natural products, eventually become a thing of the past. Just think of blueberries, raspberries, strawberries and pecans. All these crops started out as products gathered from the wild, as did all commercial mushrooms. With technology, most problems of large-scale production and commercialization can be overcome, so fewer and fewer products will eventually be harvested from wild nature in a post-modern world. For better or for worse, this is an inevitable certainty.

Two caveats. If we consider Brazil nuts as the flagship natural forest product on the global market, an age structure analysis has shown that the longest-harvested groves, those exploited for 100 years or more, are in decline due to the failure of young trees to recruit (Peres, et al., 2003 [11]). This is the logical, if not inevitable, consequence of removing all the seeds from the population for decade after decade, but on a time scale too protracted for most people to notice. Brazil nut harvesting is thus a self-limiting process that should be appropriately managed if it can ever be defined as a sustainable extractive industry.

Our second observation, based on travels through rainforests of the Americas, Africa and Asia, is that forest extractivists are among the poorest of the poor. Their livelihoods are often based on a seasonal resource that provides an irregular return, as both the volume of the offtake and market prices fluctuate wildly from year to year. The price of Brazil nuts in some years, for example, is so low that it simply doesn't pay to gather the nuts. No one dependent on such an unpredictable return can enjoy a decent life. Instead of thinking about ways to exploit harvest-sensitive natural forests "sustainably", poor rural people should be given the option of access to education that can make them employable in the regular economy.

To illustrate this point, we describe the shifting economic portfolio of the Chico Mendes Extractive Reserve (CMER) in Acre, southwestern Brazilian Amazonia, which ironically was created to slow down forest conversion into pastures following land disputes between environmental leaders like Chico Mendes and encroaching cattle ranchers. Since the mid-1990s this ~970,000-ha reserve, which is currently occupied by some 1900 households, has seen increasing diversification of income sources and land use following the collapse of rubber prices (Salisbury and Schmink 2007 [12]). Much of that alternative household-scale income, which on average remains very low (US\$281/month: Maciel, et al., 2014 [13]), has come from cattle ranching and timber extraction, mirroring the wider regional economy. For example, from 1990 to 2012 natural rubber production in Acre declined from 12,000 to 470 tons, while bovine cattle increased from 800,000 to 3,000,000 head, and native hardwood timber production from 300,000 to ~1,000,000 m^3/year. A growing cattle economy has obviously fueled much forest conversion into pastures, even inside forest reserves that had been nominally protected from deforestation. For example, Gomes (2001 [14]) recorded a 625% increase in pasture area within a single rubber estate within the CMER over a three year period (1995–1998). In fact, most previously forested rubber states at the Chico Mendes Reserve now contain cattle pastures ranging widely in size, clearly portraying a pocked landscape whereby the forest canopy of hundreds of rubber states has been perforated by exotic grasslands (Google Maps 2016 [15]).

Thus far, we have addressed intrinsic challenges to implementing sustainable development via communal management of tropical forests. There are also extrinsic obstacles that can further complicate the picture.

One is that practices can have unintended consequences. Bushmeat extraction is a prime example. All around the world, primates are targets of the bushmeat trade. Yet in most tropical forests, primates, particularly large-bodied species, constitute the most important class of seed dispersers, and their dispersal services are indispensable to the perpetuation of forest composition (Terborgh, et al., 2008 [16]). Depletion of large primate populations can then have the unintended consequence of reducing the carbon storage capacity of tropical forests via the gradual substitution of large-seeded, heavy-wooded primate-dispersed species by small-seeded, light-wooded species that are often small-bird or wind-dispersed (Peres, et al., 2016 [17]). This is, of course, a process with global implications.

Another extrinsic factor is that of enforcement, a function of the state. Communal reserves, while legally constituted, have little public visibility and, consequently, few advocates apart from the local residents, who are likely to be poor and lacking political connections. Thus, the political will required to enforce the original land use mandate is often weak or absent. Some extractive reserves established for the benefit of rubber tappers in the Brazilian Amazon, for example, have been extensively invaded by loggers and cattle ranchers who have mined high-value timber or deforested thousands of hectares without encountering official resistance (Pedlowski, et al., 2005 [18], Marques, et al., 2016 [3]). Miners, loggers and oil companies are also exerting strong political pressure to gain access to indigenous reserves, extractive reserves and even national parks in both the Brazilian and Peruvian Amazon (Finer, et al., 2014 [19], Marques and Peres 2015 [20], Pack, et al., 2016 [21]). In a world experiencing increasing resource scarcity, one wonders whether the political will to resist such pressures will be sustained.

Finally, the most important extrinsic factor influencing the management of communal reserves is that of markets. We have seen how the changing structure of markets has driven the rubber tapping economy out of the Amazon. Now, Brazil nut harvesting is under the shadow of commercial plantations of early-maturing grafted trees, and this argument could be extended to the emergence of commercial aquaculture of high-value fish species, which will compete with successful cases of community-based sustainable fishery management (Petersen et al. 2016 [22], Campos-Silva and Peres 2016 [23]). What surprises will follow next? No one can say. But what can be said with confidence is that transformative surprises are sure to occur with huge future consequences for land use across the tropics, obliterating the assumptions on which many communal reserves were established.

Conclusions

As originally conceived, communally managed reserves have the potential to benefit biodiversity by greatly increasing the area of protected forest cover in the Amazon. Some still do provide substantial biodiversity benefits. But the handwriting is on the wall. Globalization is transforming the human condition at an unprecedented rate. Populations that lived more or less sustainably in isolation from world markets are now in thrall to them as their members aspire to become modern consumers. Lured by the pull of markets, rural communities across the breadth of Amazonia desire to participate in the money economy and enjoy the fruits of technology—outboard motors, cell phones, television, and much more. Logging, cattle ranching, mining and large-scale agriculture are irreversibly transforming the economy of the region, rendering extractive reserves as obsolete as a Model T. What will the landscapes of today's communal reserves look like 20 years or 50 years from now? Unless the world comes to its senses and completely halts deforestation to avoid releasing additional gigatons of greenhouse gases into the atmosphere, we can expect that areas like the Chico Mendes Extractive Reserve will eventually resemble the surrounding unprotected landscape. Communal reserves in Amazonia were established with little thought given to the future impacts of globalization, much less the rapidity with which globalization can transform the economy of an entire region. Now, in hindsight, it is apparent that these reserves cannot be counted on to provide havens for biodiversity in perpetuity. The only policy yet devised that can ensure the perpetuation of biodiversity in perpetuity is absolute protection. May we not repeat the mistake of imagining otherwise.

Acknowledgments: J.T. and C.A.P. have no past or current grant support for research on this topic. They thank Jeff Sayer for the invitation to participate in this forum.

Author Contributions: J.T. and C.A.P. jointly wrote the article.

Conflicts of Interest: The authors declare no conflict of interest.

References

1. Schmitt, C.B.; Burgess, N.D.; Coad, L.; Belokurov, A.; Besançon, C.; Boisrobert, L.; Campbell, A.; Fish, L.; Gliddon, D.; Humphries, K.; et al. Global analysis of the protection status of the world's forests. *Biol. Conserv.* **2009**, *142*, 2122–2130. [CrossRef]

2. Peres, C.A. Conservation in sustainable-use tropical forest reserves. *Conserv. Biol.* **2011**, *25*, 1124–1129. [CrossRef] [PubMed]

3. De Marques, A.A.B.; Schneider, M.; Peres, C.A. Human population and socioeconomic modulators of conservation performance in 788 amazonian and atlantic forest reserves. *PeerJ* **2016**, *4*, e2206. [CrossRef] [PubMed]

4. RAISG. Amazonia: Protected areas and indigenous territories map. Available online: www.raisg. socioambiantal.org (accessed on 22 February 2017).

5. Ehrlich, P.R. Human impact: The ethics of I=PAT. *Ethics Sci. Environ. Polit.* **2014**, *14*, 11–18. [CrossRef]

6. Joppa, L.N.; Loarie, S.R.; Pimm, S.L. On population growth near protected areas. *PLoS ONE* **2009**, *4*, e4279. [CrossRef] [PubMed]

7. Soehartono, T.; Newton, A.C. The gaharu trade in indonesia: Is it sustainable? *Econ. Bot.* **2002**, *56*, 271–284. [CrossRef]

8. Symington, M.M. Sex ratio and maternal rank in wild spider monkeys: When daughters disperse. *Behav. Ecol. Sociobiol.* **1987**, *20*, 421–425. [CrossRef]

9. Richardson, V.A.; Peres, C.A. Temporal decay in timber species composition and value in Amazonian logging concessions. *PLoS ONE* **2016**, *11*, e0159035. [CrossRef] [PubMed]

10. Parry, L.; Day, B.; Amaral, S.; Peres, C.A. Drivers of rural exodus from amazonian headwaters. *Popul. Environ.* **2010**, *32*, 137–176. [CrossRef]

11. Peres, C.A.; Baider, C.; Zuidema, P.A.; Wadt, L.H.O.; Kainer, K.A.; Gomes-Silva, D.A.P.; Salomão, R.P.; Simões, L.L.; Franciosi, E.R.N.; Cornejo Valverde, F.; et al. Demographic threats to the sustainability of brazil nut exploitation. *Science* **2003**, *302*, 2112–2114. [CrossRef] [PubMed]

12. Salisbury, D.S.; Schmink, M. Cows versus rubber: Changing livelihoods among amazonian extractivists. *Geoforum* **2007**, *38*, 1233–1249. [CrossRef]

13. Maciel, R.C.G.; Cavalcante-Filho, P.G.; Souza, E.F. Distribuição de renda e pobreza na floresta amazônica: Um estudo a partir da reserva extrativista (resex) chico mendes. *Rev. Estud. Soc.* **2014**, *16*, 136–153. [CrossRef]

14. Gomes, C.V.A. Dynamics of Land Use in an Amazonian Extractive Reserve: The Case of the Chico Mendes Extractive Reserve in Acre, Brazil. Master's Thesis, University of Florida, Gainesville, FL, USA, 2001.

15. Google Image. Available online: https://www.google.co.uk/maps/@-10.6445753,-69.4177854,35931m/data=!3m1!1e3?hl=en (accessed on 22 February 2017).

16. Terborgh, J.; Nuñez-Iturri, G.; Pitman, N.C.A.; Valverde, F.H.C.; Alvarez, P.; Swamy, V.; Pringle, E.G.; Paine, C.E.T. Tree recruitment in an empty forest. *Ecology* **2008**, *89*, 1757–1768. [CrossRef] [PubMed]

17. Peres, C.A.; Emilio, T.; Schietti, J.; Desmoulière, S.J.M.; Levi, T. Dispersal limitation induces long-term biomass collapse in overhunted Amazonian forests. *Proc. Natl. Acad. Sci. USA* **2016**, *113*, 892–897. [CrossRef] [PubMed]

18. Pedlowski, M.A.; Matricardi, E.A.T.; Skole, D.; Cameron, S.R.; Chomentowski, W.; Fernandes, C.; Lisboa, A. Conservation units: A new deforestation frontier in the amazonian state of Rondônia, Brazil. *Environ. Conserv.* **2005**, *32*, 149–155. [CrossRef]

19. Finer, M.; Jenkins, C.N.; Sky, M.A.B.; Pine, J. Logging concessions enable illegal logging crisis in the Peruvian Amazon. *Sci. Rep.* **2014**, *4*, 4719. [CrossRef] [PubMed]

20. De Marques, A.A.B.; Peres, C.A. Pervasive legal threats to protected areas in Brazil. *Oryx* **2015**, *49*, 25–29. [CrossRef]

21. Pack, S.M.; Ferreira, M.N.; Krithivasan, R.; Murrow, J.; Bernard, E.; Mascia, M.B. Protected area downgrading, downsizing, and degazettement (PADDD) in the Amazon. *Biol. Conserv.* **2016**, *197*, 32–39. [CrossRef]

22. Petersen, T.A.; Brum, S.M.; Rossoni, F.; Silveira, G.F.V.; Castello, L. Recovery of arapaima sp. Populations by community-based management in floodplains of the Purus River, Amazon. *J. Fish Biol.* **2016**, *89*, 241–248. [CrossRef] [PubMed]

23. Campos-Silva, J.V.; Peres, C.A. Community-based management induces rapid recovery of a high-value tropical freshwater fishery. *Sci. Rep.* **2016**, *6*, 34745. [CrossRef] [PubMed]

![land logo] *land*

MDPI

Article

Can Community Forests Be Compatible With Biodiversity Conservation in Indonesia?

Agni Klintuni Boedhihartono [1,2]

[1] Centre for Tropical Environmental and Sustainability Science, James Cook University, Cairns Campus, Qld 4870, Australia; agni.boedhihartono@jcu.edu.au
[2] Tanah Air Beta, Batu Karu, Tabanan, Bali 82152, Indonesia

Academic Editors: Jeffrey Sayer and Chris Margules
Received: 19 January 2017; Accepted: 5 March 2017; Published: 14 March 2017

Abstract: Forest lands in Indonesia are classified as state lands and subject to management under agreements allocated by the Ministry of Environment and Forestry. There has been a long-standing tension between the ministry and local communities who argue that they have traditionally managed large areas of forest and should be allowed to continue to do so. A series of recent legal and administrative decisions are now paving the way for the allocation of forests to local communities. There is a hypothesis that the communities will protect the forests against industrial conversion and that they will also conserve biodiversity. This hypothesis needs to be closely examined. Conservation of biodiversity and management for local benefits are two different and potentially conflicting objectives. This paper reviews examples of forests managed by local communities in Indonesia and concludes that there is very limited information available on the conservation of natural biodiversity in these forests. I conclude that more information is needed on the status of biodiversity in community managed forests. When forests are allocated for local management, special measures need to be in place to ensure that biodiversity values are monitored and maintained.

Keywords: community forest management; biodiversity conservation; indigenous forest management; community conservation; Indonesian forests

1. Introduction

Indonesia is experiencing major changes in land allocation and tenure systems, especially in relation to forests [1]. The Constitutional Court decision No.35/PUU-X/2012 announced in 2013 was a historical landmark recognizing indigenous people's rights over their traditional forests. Until this decision was taken most natural forests were legally under state jurisdiction. The 1945 Indonesian Constitution, Article 33, declares that the State has control over all earth, water, and airspace [2]. The Constitutional Court decision No.35/PUU-X/2012 was consistent with the Rio+20 declaration on environment and development, which recognized that indigenous customary law has a vital role to play in environmental management and development because it is inspired by traditional knowledge and practices. Recent statements by the Minister of Environment and Forestry have indicated that up to 30% of Indonesia's forests could be transferred or returned to local community ownership.

AMAN (*Aliansi Masyarakat Adat Nusantara*) or the Indigenous Peoples' Alliance of the Archipelago and other non-governmental human rights organizations operating in Indonesia consider the Constitutional Court decision to be an entry point for restitution of customary rights over indigenous territories, especially indigenous forests. There are numerous *masyarakat adat* (communities that still practice traditional customary law) all over Indonesia who will now be entitled to claim that they are "indigenous" and may exploit the constitutional court ruling to obtain the release of their customary forests from designation as state forest. However, the constitutional court ruling requires that indigenous peoples be recognized by local regulations. In order to obtain this status the communities have to demonstrate that they are part of a *masyarakat adat* in their daily lives. A parallel movement is seeking acknowledgement of *'wilayah adat'* (indigenous territories) as an additional form of agrarian right residing in a separate legal arrangement [3]. When the concept of customary rights (*hak ulayat*) is being discussed as part of new regulations made by the Land Management Agency (*BPN*), each *wilayah adat* would be subject to different conditions and interpretations of rights. Customary land, with customary rights is an abstract concept describing the authority of a society and the range of customary law that applies. Indigenous territories would include land, water, air, and the customary rights attached to them. We need to understand the richness of culture and communal rights in Indonesia, but also recognize the variation amongst different communities sharing the same landscape.

The movement towards indigenous forest control has been driven by concern that corporate land grabs, especially by estate crop companies, are depriving people of their forests and land. There is a widely held view that the Ministry of Environment and Forestry has failed to protect forests against corporate interests. Laws to prevent forest clearance are weakly enforced or are blatantly ignored by elites and government patronage systems. The provision of local rights is expected to strengthen the hand of local people in defending their forests. There are moves to allocate both collective rights and private individual rights to forest lands.

The issue of collective rights or individual rights has been the object of significant discussions in many countries. The tension between collective rights and individual rights is often manifest during processes of political decentralization and attribution of local autonomy [4]. There are differing views on whether individuals are more likely to protect their forests than communities. In some cases individuals simply sell their land, whereas under community tenure this is much less likely. Brosius et al. [4] discuss the consequences of recognizing community autonomy. Brosius has observed that "When 'natives' become privileged, are other social groups marginalized? What space is there for mobility, migration, and the movements of both rural and urban poor?"

This paper is motivated by concern that the process of attributing both local communal rights and private rights is moving ahead rapidly and that adequate checks and balances to ensure forest values are protected, are not in place. Claims in the international policy discourse that biodiversity outcomes in community managed forests are superior to those of state managed protected areas may be true, but are rarely supported by evidence. Local elites can be as corrupt as the central government [5,6]. In this study, I report on observations in many parts of Indonesia over three decades of field work where I have observed the different conditions under which local people exercise protection of forest lands. My observations are that there is a very high level of diversity in local conditions and that proceeding on the assumption that "one size fits all" in local forest management is a mistake. Opportunities and threats to forests are highly context specific and a greater diversity of approaches to localizing forest management is needed. I will focus particularly on the implications of the current wave of decentralization for the conservation of Indonesia's unique biodiversity.

The issue of community forest management in Indonesia has been somewhat confused by a general failure in policy circles to distinguish between agro-forests that have been under de facto local management for generations, and more remote near-natural forests that have only been lightly exploited for non-timber products by local communities. Very extensive areas of forest in Indonesia, especially in the Greater Sunda Islands, have been progressively enriched with fruit, resin, and timber tree species over long periods of time [7,8]. These heavily modified forests have often been viewed by outsiders as natural forests and were often classified by the Ministry of Forestry as production or protection forests. Government classification restricted the use of the forests by traditional owners, and this led to tensions between these communities and the Ministry of Forestry. On 30 December 2016, President Joko Widodo announced that 12.7 million ha of forests would be allocated to local communities. He announced a decree giving "Recognition of Customary Forests" to nine Customary Communities (*Masyarakat Hukum Adat*) located in Sumatra, Java, and Sulawesi. What is unclear is whether these nine community forests and the 12.7 million ha proposed for future allocation to communities are predominantly in heavily influenced agro-forests, or whether they include more natural forests in remoter areas.

The President stated that "conservation forests whose status has changed into customary forests or rights forests, must maintain the function of conservation. It is not permitted to be traded and it should not change its function" [9]

2. Why Do We Need Biodiversity Conservation?

Before discussing whether local communities will be effective in conserving forests, I would like to discuss why societies care about biodiversity. Ehrlich and Wilson [10] postulate three reasons: "Firstly for ethical and aesthetic values, as *Homo sapiens* we have the moral responsibility to protect living companions in the universe; second to perpetuate the enormous direct benefits that people derive from biodiversity in the form of foods, medicines, industrial products, crops; third to keep natural ecosystems functioning when diverse species are key working parts of ecosystems. Biodiversity plays a critical role in ecosystem services". However, some traditional and indigenous peoples might find Ehrlich and Wilson's arguments somewhat irrelevant to their own realities—the rural poor might focus on the instrumental values that they derive from biodiversity. If we ask an elder or an indigenous person why they care about biodiversity, they will probably talk about medicinal plants, the flowers, and the fruits they use in rituals, the birds and other animals they encounter in the forests which they use in ceremonies to contact the spiritual world, and they would also value the biodiversity that contributes to their diets. Tribes living in the forests on their ancestral land regard biodiversity as contributing an integral role in the survival and sustainability of the forest ecosystem and its wildlife [11]. Traditional societies worship spirits, gods and goddesses, sacred sites, and sacred forests. Traditional people have a symbiotic relationship with nature. Traditional ecological knowledge is embedded in the complexity of peoples' lives. Local cultures and their institutions have rules, taboos, sanctions, rituals, and ceremonies, and make sacrifices which have significance for them. Respect, reciprocity, and humility are cultural values which are part of social mechanisms underpinning traditional practices. Traditional knowledge and practices are central to local systems of management that in many situations can still work sustainably in contemporary societies [12]. Traditional practices operate within complex systems which are continuously changing. As the world's population grows, people's cultures and ways of life are evolving. Development is increasing pressure on natural resources systems. Management systems have to adapt to these changing contexts, and maintaining the biological diversity that underpins the systems and their management becomes more important. These local concepts and values of biodiversity may differ significantly from the values of biodiversity held by international conservation organizations, values that are enshrined in the clauses of the Convention for the Conservation of Biological Diversity.

Environmental concerns are now central to political agendas around the world. Global mechanisms are emerging to avoid environmental degradation and to protect habitats and endangered species of plants and animals. Environmental conservation is now prominent in development strategies. We require nature to be an essential element of development [13].

Some conservation organizations have been portrayed as giving nature conservation priority over indigenous rights to self-determination [14–16]. A generation of international and national Non-Governmental Organizations (NGOs) are now seeking participation of communities in managing their forest areas and other natural resources [16,17]. Landscape approaches have emerged to provide a mechanism for balancing the interests of diverse stakeholders in managing forests and other natural resources [18], and these integrated approaches have clear relevance to local forest management.

I contend that more attention needs to be given to questions of why we conserve biodiversity. What do we wish to conserve and for whom [19]? The people who will feel the impact of conservation programs need to be involved in designing and operating them, otherwise they are unlikely to value or support them. However, merely participating in conservation programs is clearly not enough. Conservation programs have to confront the real interests and values of the communities that they impact upon. Conservation interventions can result in some winners and some losers—we need to understand who will lose and who will gain, and to put in place mechanisms to ensure equity in these processes.

For all of these reasons, community based conservation is not a simple recipe that can be applied indiscriminately in any location. Local contexts are highly diverse and societies are very heterogeneous. In any situation there will inevitably be inequitable distribution of rights and responsibilities for natural resources. We need to be critical in assessing "who actually conserves" and what do they conserve against? Berkes et al. [20] points out that "our definition of conservation is western-centric". Community-based conservation programs have to recognize the diversity of the communities that they deal with and what it means to the local beneficiaries. Kumar [21] has observed that "Community-based conservation is here to stay. The question is how a community's involvement can be made effective. Protection of biodiversity must be based on a wide range of approaches in order to develop a shared understanding of compatible conservation and development goals at various levels in societies" [21].

3. Landscapes in Transition-Development and Decentralization

Indonesia and other developing countries have rapidly expanding economies. The populations of these countries are growing and this is placing more pressure on natural resources and forests. Some Pacific islands [13] have demonstrated stark conflicts between economic development and traditional values. Communal well-being has been in conflict with individual ambition. The transition from a subsistence system to a cash economy has created deep tensions.

Clarke et al. [13] noted that "These communities developed sustained-yield systems of agriculture, agroforestry, and reef use that still operate productively today, but are in danger of disappearing in the face of changing technological, social, and economic conditions." I have observed exactly these same tensions in rural Indonesia as the country undergoes unprecedented levels of integration with the global economy.

Diamond et al. [22] dismissed the notion that pre-industrial societies lived in harmony with nature and discussed "The Environmentalist Myth". When the carrying capacity of an island or an area has been exceeded then civilizations collapsed. The rising global population threatens ecosystems, and harmony between biodiversity and people is difficult to achieve. Again, in forest areas of Indonesia I have observed exemplary practices by indigenous communities in managing nature, but I have also seen cases of flagrant disregard for sustainability by people pursuing their individual interests.

There is a diversity of conservation behavior amongst traditional societies, and nowhere is this greater than in Indonesia. Chapman et al. [23] has argued that "there is an urgent need to improve understanding of conservation attitudes in the Third World because of the increasing rate of resource depletion that is now occurring in the countries involved". Chapman et al. [23] has argued that whilst "conservation practices by traditional societies have received much attention from research workers, the fact that some practices are intentional and others inadvertent has been largely ignored". In the popular discourse on local forest management in Indonesia there is recognition of this diversity, but there are few empirical studies of how this diversity of management systems operates in the forest.

Duncan et al. [24] examined the growing numbers of NGOs and other institutions in Indonesia engaging with indigenous rights movements in the hope that decentralization processes would allow ethnic minorities to retain or regain control over natural resources through local-level political initiatives. Some ethnic minorities see decentralization as an opportunity to return to local land tenure customs and resource management systems that were opposed by the national government during the Soeharto era. However, the decentralization process has encouraged district governments to over-exploit natural resources in an attempt to generate more income for their district. Minority communities have sometimes suffered negative consequences from decentralization processes, as local governments disregard their land rights in attempts to raise income. Local governments often pursued the same rent seeking behavior as the Soeharto Regime [24–26].

Indigenous ethnic minorities in Indonesia are impacted by the implementation of decentralization and regional autonomy policies. Decentralization is supposed to better protect the interests of ethnic minorities and other marginal groups within the state as local communities gain more control over their own affairs [27]. However, others contend that decentralization allows local elites to get more benefits while still excluding ethnic minorities and other vulnerable populations from the political process [28,29], and women's concerns seem to be largely ignored in these debates.

Brosius et al. [4] expressed concern at the proliferation of movements and agendas promoting the "indigenous paradigm" and argued that donor institutions and government agencies need to be careful in assessing claims and counter-claims, and must base their decisions on evidence. The international discourse amongst human rights advocates, international and national NGOs, and other activists may sound compelling, but Brosius suggests a closer examination of the extent to which the advocates of local resource management are actually reflecting local concerns and needs. International and local organizations have been advocating the establishment of Community-Based Natural Resources Management (CBNRM) programs in different parts of the world, but the results are perceived differently by advocates depending on whether they come from a conservation institution, development organization, or they genuinely represent indigenous people.

Local wisdom, traditional knowledge and values of indigenous groups are important when involving communities in natural resources management. However, for many indigenous groups the term conservation is seen as a concept "exclusively for the elites" or westerners. For many indigenous peoples living in poverty, short-term survival is their daily preoccupation. Activists and human rights groups championing indigenous access to forests argue that people have used the forests for their subsistence needs for generations. They accuse conservationists of taking forests to establish protected areas or national parks [14,30]. These processes are unfolding in landscapes that are subject to rapid change. People's needs and expectations are evolving. The dilemmas created by conflicting demands on land need to be resolved through landscape scale processes that allow all stakeholder voices to be heard [31].

4. The Importance of Traditional Knowledge in Conservation

Traditional knowledge, as a way of knowing, is similar to western science in that it is based on an accumulation of observations, but it is different from science in some fundamental ways [12]. Scientists take notes and record data as they seek to understand causal relationship. These studies can be replicated at different times, by different people or in other places. Traditional people accumulate knowledge in a more experiential and informal way—they learn from doing, and over time they will adapt and retain best practices and discard practices that do not serve their purpose. They do not conduct controlled experiments.

Mainstream conservationists have often focused on conserving biodiversity and are sometimes reluctant to accept that indigenous people are part of the landscape where conservation is occurring. Only recently have significant numbers of conservationists switched to the belief that the world's biodiversity is found and will continue to be found in landscapes occupied by people [17].

Brosius et al. [4] stated that: "Community-based natural resource management is imagined differently by different advocates. Conservationists, both indigenous and foreign, hope to involve local people in achieving transnational conservation and resource management goals as a means of protecting biological diversity and habitat integrity" [32–34]. However, the empirical evidence for community based management delivering both local benefits and broader environmental benefits is quite mixed.

Traditional ecological knowledge is defined as a cumulative body of knowledge, practice, and belief evolving by adaptive processes and handed down through generations by cultural transmission, about the relationship of living beings, including humans, with one another and with their environment [35]. Traditional ecological knowledge is accumulated over many generations through trial and error. Not every indigenous group has such a body of traditional ecological knowledge, but many communities do retain a body of such knowledge [35]. My experience in Indonesia reflects this diversity in the retention and value of traditional ecological knowledge. My studies of the Punan in Kalimantan in the 1990s showed significant retention of sophisticated ecological knowledge, but in the decades that have elapsed since that time many Punan have moved to towns and are shifting towards more mainstream livelihood systems [36–38].

Berkes et al. [12] have emphasized that recognition of the importance of traditional ecological knowledge has been growing and it is now widely accepted that such knowledge can contribute to the conservation of biodiversity [39], rare species [40], protected areas [41], ecological processes [42], and to sustainable resource use in general [43]. Berkes et al. [20] again highlighted "the use of local and traditional ecological knowledge as a mechanism for co-management and empowerment". This is true in the case of Indonesia, but again the effectiveness of traditional ecological knowledge systems is highly variable according to where we are in the archipelago.

There are numerous examples where traditional knowledge practices have melded with western ideas of conservation resulting in successful action drawing on ideas from different cultures and producing multi-disciplinary science-based management outcomes [44,45]. Several conservation NGOs have successfully worked with local knowledge systems in Indonesia to achieve conservation outcomes, while other conservation organizations have been less receptive to local perspectives and knowledge.

The system of *adat* (customary law) and religious belief systems clearly has an important role to play in community-based conservation initiatives in Indonesia [21,36,46]. Wadley and Colfer [47] have shown that Iban agroforestry systems and their land use practices may be important for local economic purposes, but they may also be valuable in promoting and enhancing the more global goals of biodiversity conservation. Indigenous peoples with a historical continuity of resource-use practices often possess a broad knowledge base on the behavior of complex ecological systems in their own localities. This knowledge has accumulated through a long series of observations transmitted from generation to generation. Such "diachronic" observations can be of great value and complement the "synchronic" observations on which western science is based [39].

Brosius et al. [48] argues that " ... indigenous k*nowledge* is generally applied to discussions of indigenous understandings of the natural world: systems of classification, how various societies recognize or interpret natural processes, what such groups know about the resources they exploit, and so forth ... ".

Berkes et al. [12,43] again emphasize that "traditional knowledge is a knowledge–practice–belief complex, evolving by adaptive processes and handed down through generations by cultural transmission, it is about the relationship of living beings (including humans) with one another and with their environment". Gadgil et al. [39] argue that traditional ecological knowledge is an attribute of societies with historical continuity in resource use practice. My work with indigenous communities in the remote forests of Indonesia strongly supports these observations.

As the world changes, societies' priorities change. *Adat* traditions and cultures are living systems and they adapt and change. A lot of communities abandon traditional local religions and convert to the major world religions, and young people no longer see the relevance of rituals or ceremonies in their lives [36]. Parents send their children to school in towns. When they finish school they no longer want to become farmers, fisherfolks, or hunter-gatherers. Transfer of knowledge and values changes over time. Throughout Indonesia I observe that many young people are changing, adapting, and revising their traditions and communal values to conform to their new environments.

Dahl et al. [49] has observed that children who are no longer educated in the family or the tribe, but in schools with western-style education, tend to change and adapt cultures to their new way of life. Traditional patterns of social organization for collective action are disrupted, making it difficult to continue group occupations such as collective fishing or the irrigated cultivation of rice or taro. New occupations in towns, mines, or commercial agriculture attract the most able and the people really in need of jobs to fulfil their family needs, and reduce dependence on traditional subsistence activities. Traditional knowledge no longer passes automatically from parent to child. Even where subsistence activities have continued, new technologies replace old ones and old knowledge seems superfluous or redundant. The technological fix becomes a temptation for all societies.

5. Can Community Conservation work in Indonesia?

Conservation in a landscape or seascape where indigenous or *adat* communities and local communities have strong commitments and strong informal institutions does still work. *Awiq-awiq* and *Sasi* systems are traditional systems of managing natural resources to prevent over-harvesting. *Awiq-awiq* and *Sasi* are still widely respected in the islands of eastern Indonesia. However, local context will determine which approaches to conservation will work in different landscapes or seascapes.

In Indonesia, forests are being given back to local communities to manage. These communities are not static, they are constantly changing. Younger generations are abandoning rural communities to move into cities to have a "better life". They leave in search of salaried employment. They aspire to own cellphones, T-shirts, motorbikes, and many other material benefits that the city offers [38]. The communities that may have traditionally managed the forests may no longer exist in their original form. *Adat* systems are not locked in time, they are evolving in multiple ways, and it cannot be assumed that they will function as they did in the past.

The people left behind in rural villages often remain poor and have to struggle to meet their basic needs. Outside investors provide opportunities, but also pose threats. Conservation of their natural environment may or may not remain important in their livelihoods. There is a deepening gap between the elders who still retain traditional knowledge and values and the younger generation who no longer practice taboos or restrictions on exploitation of forest resources and who aspire to obtain immediate material benefits. Thus, elders' concerns to conserve the environment and the forests resources for their children and grandchildren is there, but more powerful actors now have access to what used to belong to the communities. If local people do not take the forest products, then external actors will. This is a strong disincentive for conservation.

Taboos are often mechanisms for the protection of species and habitats and may work in contemporary society where they complement other social rules and sanctions, rooted in traditional belief systems [43]. Taboos may serve a social function in the management of natural resources, but in Indonesia a lot of traditional societies no longer practice those taboos or restrictions which support biodiversity conservation.

Colding et al. [40] have argued that taboos represent unwritten social rules that regulate human behavior. Such constraints may not only govern human social life, but also may affect, and sometimes even directly manage, many components of the local natural environment. Whatever the reason for such constraints, taboos may, at least locally, play a major role in the conservation of natural resources, species, and ecosystems [23,39,41]. There are critics who view the practice of taboos as irrational and a hindrance toward development [50], who dismiss any ecological reasons behind them [51], or who argue that the taboos may not be adhered to by some groups and, consequently, may be of no value in nature conservation [52]. Numerous situations persist in Indonesia where taboos clearly do continue to serve to protect nature, but the strength of these taboos is declining.

My own observations studying taboos that persist amongst remote communities are consistent with those of Berkes et al. [53] who describe social restraints, such as taboos, that lead to indigenous biological conservation. These restraints include providing total protection to some biological communities, habitat patches, and certain selected species, as well as protection of other species during critical stages of their life history.

The literature on the potential for biodiversity outcomes to be achieved better through the use of traditional knowledge and by placing management responsibility in the hands of local and traditional people is rich and diverse. There are claims and counter-claims often poorly supported by empirical evidence. There are many general assertions used by campaigning groups which fail to recognize the extreme importance of local context in determining approaches that will work. The following accounts are based upon three decades of field work with different traditional groups in Indonesia and illustrate the complexity of the situation in the field. Figure 1 shows the locations of the *adat* communities discussed.

Figure 1. Locations of the *adat* communities discussed.

6. Examples of *Adat* Communities' Relations with Their Forests

6.1. The Baduy of West Java

The Baduy people of West Java are a community who are still strongly dependent on their *adat* and traditional belief systems and have retained complex land management practices. The Baduy inhabit an area in the mountains of West Java less than 200 Km from Jakarta, the capital of Indonesia. The Baduy have two communities. The "Inner Baduy" wear white garments and are largely isolated from surrounding communities. The "Outer Baduy" wear black garments and have a greater degree of contact with outsiders. The Baduy practice norms and beliefs that are articulated in the form of a code of conduct and they have taboos that mediate their daily behavior. Baduy traditions forbid changes to their landscape and the ecosystem. Therefore, levelling the land to make houses or for irrigation, dams, and waterholes is taboo.

Cosmological beliefs inherited from ancient eras concerning nature and the earth are the main source of Baduy norms, beliefs, and culture. They believe that land and natural resources have a spirit. They glorify their homeland and do not allow anything to disturb the ecosystem. They protect sacred forests where they believe their ancestors reside by offering "sacred" swidden rice to maintain and reinforce their links with their ancestors.

The Baduy have strict codes of conduct with supporting taboos which are implemented in farming, extracting natural resources, and other nature-related activities. To protect their norms, beliefs, culture, and behavior from outside influences, some taboos are additionally imposed, appropriate to the cultural context. Although their customary law is complex and open to multiple interpretations, there are ecological justifications behind those customs. All codes of conduct and related taboos which contribute to customary law have environmental implications and can mediate ecological strategies. To regulate their customs, the Baduy community has a unique traditional governance system which still functions strongly today. To enforce customary law and control violation of taboos, self-control mechanisms (sin), social control, and punishment are imposed [54]. The net result is that areas of natural forests are conserved within the Baduy landscape and as far as can be determined the natural biodiversity of these forests is protected.

"The *Kawalu* ritual performed by Baduy people is one of many mechanisms which cleanse the community and which punish people who violate taboos by imposing them to leave their community. In this way, the integrity of the Baduy community is maintained" [54]. Land is under communal control and held by the *Pu'un*, who is the supreme leader, but individual access to use, claim, or control certain land and natural resources is allowed to support livelihoods. There is evidence for the erosion of some traditions and practices by many Baduy people, and some have become "contaminated" by modern lifestyles. They then have to leave the Inner Baduy village and move to the Outer Baduy village as part of their punishment. The trend towards gradual erosion of customary law still continues in the Baduy community. Generally, both the Inner Baduy and the Outer Baduy people still believe and observe their *adat* and belief system (*Sunda Wiwitan*). The Inner Baduy hamlets have a structure consisting of *Pu'un*, *Girang Serat*, and *Baresan*, and they form the core of the traditional council of the Baduy community. As with many indigenous people, the Baduy face a dynamic and changing world and so they will certainly have to adapt to and accept environmental, social, economic, and political changes. The main question is how long the Baduy survive with their unique culture and how durable their *adat* institutions will be as external influences increase and when the pressure of globalization is even greater in the future.

The present behavior of the Baduy does ensure the maintenance of forest cover. They are effective in preventing outsiders from clearing or otherwise exploiting their forest. The forest habitat is therefore protected. Detailed biodiversity surveys have not been conducted in Baduy areas, so the extent to which the full range of biodiversity in the forests is protected-is not known.

6.2. The Iban of West Kalimantan

Indigenous people of Kalimantan were called "Dayak" (people from the interior) by the Dutch, but they use more specific terms to describe their ethnicity in their own languages. Many different Dayak groups exist and include the Iban, Kenyah, Kayan, Ngaju, Lundayeh, Merap, Embaloh, etc. Dayak communities in Kalimantan are known for their traditional belief systems, taboos, norms, and traditional knowledge of their environment and especially of medicinal plants [36,47]. The Iban do not harvest excessive quantities of natural products, and they practice strict resource management systems [55]. The sustainability of these Iban forests management practices is well documented in a study by Dennis et al. [56]

A study of the conservation of orang utan populations in West Kalimantan has shown that conservation should not focus exclusively on the single species, but rather on the need to maintain social and natural capital, cultural diversity, and ecological functions at various institutional levels and across geographical scales [57]. Some Dayak groups still practice taboos, and have restrictions on the hunting of orang utans. The Iban community believe that their ancestors were helped by the orang utan during a tribal war, so they cannot hunt them. Some younger Iban who have converted to Christianity do not retain the same values as their grandparents or parents. As informal institutions weaken, development becomes a threat to the community norms and practices.

Social movements are important to create transformative cross-scale communication that could recouple global, national, and local society [58]. Kalimantan is under pressure from development and its people are increasingly integrated into the cash economy [26,59]. As with other tropical forest areas, the people have to deal with the arrival of industrial plantations and mines. Some of these indigenous people struggle to adapt to new technologies and rapidly changing lifestyles. Resource extraction driven by international investors and national governments is transforming landscapes. Local administrations require resources for the growth of the economy and the situation is further complicated by the arrival of migrants attracted by opportunities of employment in agriculture and agro-industries. The changes in the lives and institutions of the Dayak communities are widely documented by Tsing [26].

Dayak sustainable resource management faces two problems: modernization, which has altered Dayak lifestyles over the past several decades, and the Indonesian central government's attempts to control access to resources on Dayak lands [26,58]. I spent several months with the Iban in their longhouses in the Kapuas Hulu in 1997 and observed that many traditional practices were still maintained. On visits to the area between 2012 to 2015, I found that the Iban were still very interested in the idea of conservation of the forests around the Betung Kerihun National Park and the upper Kapuas areas, but pressure of development from migrants and estate crop industries in the region are increasing, and people fear that they will be marginalized if they do not integrate into these industries and become less dependent on the forests. The degree to which the Iban continue their traditional conservation practices is declining and this decline appears likely to accelerate in the future.

6.3. The Mentawai of the Siberut Island off West Sumatra

I spent several months living with the Mentawai in 1999 and 2000 and observed that they retained many of their traditional natural resource management practices. More recently, Quinten et al. [60] studied hunting of four endemic primates of Siberut Island by the Mentawai and examined the attitudes of indigenous inhabitants to resources utilization. Quinten et al. [60] assessed the scale and impact of hunting of primate populations to determine if hunting could be sustainable. They concluded that present levels of hunting could be sustainable, but that when hunting combines with other forms of disturbance such as logging and habitat loss the outcome is likely to be unsustainable [60]. Mentawai has one of the world's highest levels of primate endemism [61] and is a critically important area for global primate conservation. The main threats to primates are habitat loss, land use change, and hunting. Logging, land clearance, and agricultural expansion driven by immigrants from other parts of Indonesia are leading to significant loss of intact forests across the Mentawai islands [60,62]. When I lived with the Mentawai in the 1990s, primate hunting and the consumption of primate meat were important parts of rituals and ceremonies associated with the belief system of the Mentawai people.

The *Arat Sabulungan* (Mentawai traditional belief system) holds that every living thing and any moving object including waterfalls, rivers, and the wind together with stationary objects such as rocks and trees and natural phenomena such as rainbows, thunder, and lightning, that each of these things has a soul (*simagre*). Delfi [63] explained how the Mentawai believe that human interventions in nature and other objects have consequences which could lead to imbalance or disturbance. They believe that it is important to restore the balance through ceremonies which attempt to heal and recover what was lost. These traditional cultural practices were very much in evidence during my stays with the Mentawai in the 1990s. However, the distinctive way of life and elaborate religious ceremonies centered on the *umah*, ceremonial house, are under threat from the Indonesian government which wishes to 'civilize' the Sakuddei people [64,65].

Until recently, the traditional Mentawai faith of *Arat Sabulungan* was not recognized as a proper religion by the government of Indonesia. The Mentawai were encouraged to embrace other religions if they wanted to advance and become "modern". Threats and harsh treatment by the Indonesian military were used to discourage people who were practicing *Arat Sabulungan* [65]. The Mentawai still practice their traditional beliefs even though they have officially adopted Christianity or Islam [66]. The *kerei* (medicine man or shamans) still lead ceremonies and heal the sick.

The Indonesian government has consistently tried to develop remote areas of Indonesia by launching "modernization programs" for *masyarakat terasing*. This term *masyarakat terasing* translates as "isolated communities" and is used pejoratively on the assumption that these groups are isolated from 'modern Indonesian culture' and do not contribute to the unity of the nation [25,67]. After independence, Indonesia's main goals were national unity and cultural adaptation. Isolated communities had to adapt their agriculture, religion, and education [66]. On Siberut Island, as with many other islands, these government actions had strong impacts on the people, especially the younger generations. Traditional Mentawaian religion was forbidden, and *kerei* were told to abandon their ceremonial objects and withhold from further practice. Everyone had to convert to a monotheistic religion (one of the five religions that were allowed by the government: Islam, Catholicism, Protestantism, Hinduism, and Buddhism); tattoos and loin cloths were forbidden as they were considered "primitive". Since the fall of the Soeharto regime there has been a welcome change in government policies towards traditional religions and practices. Identity cards previously only allowed for recognition of one of five monotheistic religions, however, since 2016 people have been allowed to record their local religion/belief system on their identity cards. Laws allowing community management of forests also reflect this liberalization of attitudes towards ethnic minorities. This new acceptance of diverse local religions may provide conservation benefits if it allows traditional conservation practices to achieve legitimacy.

6.4. The Kajang of South Sulawesi

Between 1994 to the late 1990s I visited the Konjo speaking Kajang communities who live in Bulu Kumba district in South Sulawesi on several occasions. The *Ammatoa* customary community or *Kajang* (*Kajang Dalam* or "inner circle") communities with their traditional belief in *Patuntung* have a tradition of leading a simple and humble life. *Adat* institutions in Kajang territory are aligned with the mystical belief system of the spirit cult of the *Ammatoa* [68]. The *Ammatoa* (the *adat* leader) exercises supreme control over spiritual and moral affairs as he is the supreme leader in affairs relating to *adat* rules defining socio-religious norms and morality. He is the protector and controller of the community with respect to moral and spiritual aspects of social life [69]. The *Ammatoa* is considered to have supernatural power and people around Bulu Kumba believe that the Kajang practice mysticism. Outsiders fear the Kajang, and this is probably why forests are still intact in the area.

Whilst staying with the Kajang I observed that strict regulations applied to many aspects of Kajang behavior and that severe punishments were applied to people who transgressed these rules. For example, when a person wanted to cut a tree, he was required to plant two similar trees. Harvesting honey from the forest was only allowed for feasts and ceremonies. Taking medicine from the forest was only allowed in case of genuine illness and not for sale. The *Pasang* was the point of reference for the people and the source of religious values [69]. The people believed that the *Pasang* communicated messages from ancestors. *Pakpasang* messages from God described the relationship between the Creator and human beings. Forests are sacred places. Kajang are not allowed to change any land or the soil surface. They are required to build their houses using unprepared logs rather than sawn timber or other modern materials. Areas of natural forest were retained in the Kajang landscape.

The Kajang are one of the groups to whom communal forest rights have now been granted. The Kajang forests are home to a number of rare and endangered animals and plants endemic to Sulawesi. The award of forest rights to the Kajang should contribute to the conservation of the forests and their endemic species, but, as elsewhere, there is little empirical evidence to allow us to judge just how safe this biodiversity will be. At this stage a thorough study of the biodiversity in the area is needed to monitor biodiversity richness and landscape changes.

6.5. The Punan Tubu of East Kalimantan

Conservation of forests and biodiversity has different meanings and values for the Punan. The Punan in Malinau and the Tubu catchments are hunter-gatherers. They now face development pressures and roads are opening up to the interior of the Tubu highlands. Punan communities are no longer nomads, but mobility is still high [37]. Some Punan Tubu communities live in the nearby town of Malinau (in *Respen* Sembuak, a resettlement village that has existed since the 1970s) and other members live in remote highlands areas about a four-day trip by motor canoe from Malinau. Only recently, a road has been opened to the Tubu area which will make the pressure even greater on the natural forests in the region.

Remote Punan communities are very dependent on trade, their main income comes from the forests. Sellato [70] and Cunliffe [71] explained the importance of trade in the life of the Punan where they live in symbiosis with other Dayak groups, and they rely on *gaharu* and other Non-Timber Forest Products (NTFP) to barter for their basic needs. They provide *gaharu* (*Aquilaria* sp.), hornbill ivory, bird nests, camphor, bees wax, rhinoceros horns, resins and gums, rattan, bezoar stones, tusks and furs of wild pigs, leopards, and bears, antlers of deer, etc. in exchange for salt, tobacco, iron products, and cotton textiles that the other Dayak groups can provide. Nowadays the Punan are still dependent on *gaharu* and bird nests, but these products are harder to obtain.

Studies of the value of forests and their importance to the local communities in the Malinau landscape [72,73] provide local perspectives on the importance of biological diversity using the methods of multidisciplinary landscape assessment. The myth that Punan leave their dead family members in the forests in large ceramic jars persist. The Punan prefer not to have outsiders going into their territory. People therefore would not dare disturb sacred forest areas. Gravesites are very important locations in the forests with rich biodiversity conservation values, as these sites are usually protected and free of Non-Timber Forest Product collection. No one is allowed to go into sacred forests and therefore these are potentially valuable biodiversity conservation areas. However, the large ceramic jars which are very valuable are now often stolen. The destruction of gravesites has become a concern when concession companies open up these areas. Concession companies do not understand nor value the importance of sacred forests for the Punan [73].

Punan Tubu still consider forests as their main livelihood resources; some claim that it is their safety net. Hunter-gatherers who become sedentary tend to lose the diversity of food resources that are rich in protein and fiber [74]. Punan closer to urban areas are in better health as they are closer to health care and markets. However, some Punan from *Respen* still wander seasonally into the forest in search of forest resources which enrich their diets and provide extra income.

Forest biodiversity is important to the Punan because of their dependence on livelihood activities which include harvesting food and crop resources, and the collection of Non-Timber Forest Products. [71]. The Punan also want to integrate with the cash economy and value the advantages that development brings to their daily lives. The Punan appreciate better livelihoods and they certainly like to catch up with their Dayak neighbors who have good jobs and better agriculture systems [38]. Attempts to engage the Punan in conventional agriculture system have often failed and they may need to find systems that are more suited to their local context. The Punan will need to adapt to new conditions and integrate themselves into the cash economy to survive in a changing world. The ability of the Punan to actively conserve their biodiversity may be declining.

7. Conclusions

I have restricted my review to a small number of communities with whom I have had personal experience. They are just a few of the hundreds of communities that I might have chosen. They illustrate the fact that the elements needed for successful community management still exist in many locations, but that even in these remote areas they are rapidly being eroded by the forces of modernity. The extent to which these traditional practices can contribute to biodiversity conservation will depend upon the persistence of local management systems and the regulatory conditions that support them.

Conservation is not a uniformly understandable concept. Conservation has multiple interpretations. Biodiversity can relate to plants, wildlife, knowledge, culture, etc. With the changing world and increasing pressures of development there are inequitable distributions of rights and responsibilities for natural resources [21]. Kumar has noted that community-based conservation is here to stay. The question is how community involvement can be made effective. Conservation can work if it is undertaken by recognizing and building on what local people find important and on what matters for their livelihoods [73]. Protection of biodiversity must be based on a wide range of approaches to develop a shared understanding of compatible conservation and development goals at various scales. I emphasis the principle that conservation has different meanings and objectives for the communities who manage forests from those of international environmental negotiators. Management of forests for local communities is not necessarily going to conserve biodiversity.

Masyarakat terasing (isolated communities) in Indonesia such as Baduy, Kajang, Mentawai, some Dayak groups, Bajau, and others who live in remote parts of the archipelago are all seeking improvements in their material well-being. Many communities still have strong informal institutions, and still practice traditional belief systems. Most of these peoples would prefer that their forests should still be in good condition, but development provides many of the improvements in their lives that they seek [36]. How long will they be able to survive in a changing world given the influence of the cash economy that dominates modern societies? How can these people adapt their way of life to this pressure of development, and will they still find it worthwhile to conserve biodiversity?

In Bali, people are strongly connected to the modern economy and are experiencing growing economic development, but their traditional belief systems are still maintained. Some forested landscapes are still intact even in accessible areas such as around the Batu Karu temples just 40 km from the tourist areas of Kuta and Seminyak. People in these *Bali Aga* communities still have deep respect for their Gods, Goddesses, and Spirits. Their life is still linked to their environment in the same way as the Baduy, the Kajang, the Mentawai, and other ethnic groups. The rural Balinese still believe that misfortune will occur if they harm the environment, their forests, their water and trees. However, in these communities an increasing proportion of the people are moving out to live, work, and study elsewhere, and these people tend to be less respectful of traditional values. Even if a major part of the population of a village respects traditional conservation measures it is possible for a small minority who do not respect these values to do great harm to the environment.

Conservation of biodiversity values in locally managed forests in Indonesia will only occur if certain preconditions are met. It is unrealistic to assume that *adat* rules will be adequate to ensure conservation. *Adat* rules relate to features of the landscape that have instrumental or cultural values and much biodiversity might not have such local values. Even *adat* rules that might have protected areas of sacred forest against any form of disturbance have been weakened by over a hundred years of deliberate attempts to impose uniform cultural and social values upon the Indonesian people [25]. The communities whose traditional behavior is described in this paper are exceptions to the general rule—they are communities that have sought to distance themselves from modernity and have retained a higher dependence on *adat* rules than most other communities. The erosion of *adat* rules does not mean that other communities will not protect biodiversity. There are many situations where biodiversity does provide instrumental value to communities. The people around the Tangkoko reserve in North Sulawesi protect birds, the Spectral Tarsier and the Black Macaque, because they make money from visiting eco-tourists. Local people protect the display sites of birds of paradise in the Arfak Mountains in West Papua and on the island of Halmahera because visiting bird watchers contribute to the local economy. Beautiful seascapes in several Eastern Indonesian islands are starting to yield income for local people as scuba divers come to see the richness of coral reefs and marine biota that are globally rare. Many other similar examples exist throughout the archipelago.

However, many biodiversity values are not apparent to local communities. The high value attached to obscure species by international conservation interests is difficult to understand for rural people. In many cases protecting these "global public goods" values of biodiversity would only make sense to local people if they received some benefit from doing so. Payments for environmental services might encourage people to protect such species and their habitats, but up until the present such payments have been promised but have rarely materialized [75]. Understanding that incentives are multidimensional and that equity and empowerment sometimes are more important than monetary incentives are essential to make things work [20].

Decentralizing forest management to communities is desirable and communities must surely be better stewards of forest values than the government agencies that have failed so miserably to do this in recent decades. However, simply handing over control is not enough. The nature of local control, the rules under which it operates, and the institutions needed to ensure equity and justice in exercising local rights will be different in different locations. There is no "silver bullet" to ensure the success of local management. Arrangements will have to be tailored to local conditions. Biodiversity values differ in their susceptibility to change and local capacities are variable. Optimal conservation strategies may differ radically depending on the objective and the location. Good governance and strong government institutions are vital, but Indonesia has struggled to make its forest institutions work effectively. Greater recognition of *adat* institutions and traditional belief systems are needed, and decentralization should strengthen, not weaken these traditional systems. Legal and regulatory systems must be strengthened. Strong community leaders will be needed. Indonesia has the world's highest levels of biodiversity and the conservation of this unique resource will depend upon the transition to decentralized management being conducted thoughtfully and carefully and with full recognition of the potential for failure if the right match between modern and traditional cultures and practices is not achieved.

One vital step in using the decentralizing process to achieve conservation outcomes is the accumulation of evidence. More research is needed to improve our understanding of the maintenance of biodiversity in all forests areas. At present, biodiversity is being lost in forests managed by government and by communities. There is an urgent need to produce evidence on biodiversity outcomes in locally managed forests. If biodiversity is to be conserved, mechanisms must be put in place to assess biodiversity outcomes in all government, corporate, and community managed forests. We have argued elsewhere for more effective ways of measuring the performance of complex forest landscapes [76]. The extent of natural forest cover, the protection of key species, and the ability of communities to protect their forests against outsiders must all be subject to measurement. Only by doing this will we have the empirical knowledge needed to justify expansion of the role of communities in biodiversity conservation.

Acknowledgments: I would like to thank my "adoptive families" in the Baduy, Mentawai, Iban, Kajang, Punan, and Bali Aga communities with whom I have stayed and who have been very patient with me during my long periods in their company. These encounters inspired this paper. I would also like to thank Jeff Sayer and two anonymous reviewers for comments on earlier versions of this paper.

Conflicts of Interest: The author declares no conflict of interests.

References

1. Riggs, R.A.; Sayer, J.; Margules, C.; Boedhihartono, A.K.; Langston, J.D.; Sutanto, H. Forest tenure and conflict in indonesia: Contested rights in Rempek village, Lombok. *Land Use Policy* **2016**, *57*, 241–249. [CrossRef]
2. Nurjaya, I.N. Adat community land rights as defined within the State Agrarian Law of Indonesia: Is it a genuine or Psuedo-legal Recognition? *US–China Law Rev.* **2011**, *8*, 380.
3. AMAN Website. Available online: https://www.aman.com/ (accessed on 29 November 2016).
4. Brosius, J.P.; Tsing, A.L.; Zerner, C. Representing communities: Histories and politics of community-based natural resource management. *Soc. Nat. Resour. Int. J.* **1998**, *11*, 157–168. [CrossRef]
5. Alatas, V.; Banerjee, A.; Hanna, R.; Olken, B.A.; Purnamasari, R.; Wai-poi, M. Does Elite Capture Matter? Local Elites and Targeted Welfare Programs in Indonesia. Available online: http://www.nber.org/papers/w18798 (accessed on 19 January 2017).
6. Lucas, A. Elite Capture and Corruption in two Villages in Bengkulu Province, Sumatra. *Hum. Ecol.* **2016**, *44*, 287–300. [CrossRef] [PubMed]
7. De Foresta, H.; Michon, G. Complex Agroforestry Systems and Conservation of Biological Diversity (II). Available online: http://agris.fao.org/agris-search/search.do?recordID=MY9205324 (accessed on 19 January 2017).
8. Michon, G.; de Foresta, H. Forest resources management and biodiversity conservation: The Indonesian agroforest model. In Proceedings of the IUCN Workshop on Biodiversity Conservation Outside Protected Areas, Madrid, Spain, 8–12 March 1994.
9. Secretariat Cabinet of Republic of Indonesia Website. Available online: http://setkab.go.id/en/home/ (accessed on 30 December 2016).
10. Ehrlich, P.R.; Wilson, E.O. Biodiversity studies: Science and policy. *Science* **1991**, *253*, 758. [CrossRef] [PubMed]
11. Madegowda, C. Traditional knowledge and conservation. *Econ. Polit. Wkly.* **2009**, *44*, 65–69.
12. Berkes, F.; Colding, J.; Folke, C. Rediscovery of traditional ecological knowledge as adaptive management. *Ecol. Appl.* **2000**, *10*, 1251–1262. [CrossRef]

13. Clarke, W.C. Learning from the past: Traditional knowledge and sustainable development. *Contemp. Pac.* **1990**, *10*, 233–253.

14. Cernea, M.; Schmidt-Soltau, K. National parks and poverty risks: Is population resettlement the solution? In Proceedings of the World Parks Congres, Durban, South Africa, 8–17 September 2003.

15. Cernea, M.M.; Schmidt-Soltau, K. Poverty risks and national parks: Policy issues in conservation and resettlement. *World Dev.* **2006**, *34*, 1808–1830. [CrossRef]

16. Colchester, M. Self-determination or environmental determinism for indigenous peoples in tropical forest conservation. *Conserv. Biol.* **2000**, *14*, 1365–1367. [CrossRef]

17. Alcorn, J.B. Indigenous peoples and conservation. *Conserv. Biol.* **1993**, *7*, 424–426. [CrossRef]

18. Sayer, J.; Sunderland, T.; Ghazoul, J.; Pfund, J.L.; Sheil, D.; Meijaard, E.; Venter, M.; Boedhihartono, A.K.; Day, M.; Garcia, C.; et al. Ten principles for a landscape approach to reconciling agriculture, conservation, and other competing land uses. *Proc. Natl. Acad. Sci. USA* **2013**, *110*, 8349–8356. [CrossRef] [PubMed]

19. Boedhihartono, A.K.; Sayer, J. Forest landscape restoration: Restoring what and for whom? In *Forest Landscape Restoration: Integrating Natural and Social Sciences*; Stanturf, J., Lamb, D., Madsen, P., Eds.; Springer: Dordrecht, The Netherlands, 2012; pp. 309–323.

20. Berkes, F. Rethinking community-based conservation. *Conserv. Biol.* **2004**, *18*, 621–630. [CrossRef]

21. Kumar, C. Whither 'community-based' conservation? *Econ. Polit. Wkly.* **2006**, *41*, 5313–5320.

22. Diamond, J.M. Archaeology: The environmentalist myth. *Nature* **1986**, *324*, 19–20. [CrossRef]

23. Chapman, M.D. Environmental influences on the development of traditional conservation in the South Pacific Region. *Environ. Conserv.* **1985**, *12*, 217–230. [CrossRef]

24. Duncan, C.R. Mixed outcomes: The impact of regional autonomy and decentralization on indigenous ethnic minorities in Indonesia. *Dev. Chang.* **2007**, *38*, 711–733. [CrossRef]

25. Li, T.M. *The Will to Improve: Governmentality, Development, and the Practice of Politics*; Duke University Press: Durham, NC, USA, 2007.

26. Tsing, A.L. *Friction: An ethnography of Global Connection*; Princeton University Press: Princeton, NJ, USA, 2011.

27. Kaimowitz, D.; Vallejos, C.; Pacheco, P.B.; Lopez, R. Municipal governments and forest management in lowland bolivia. *J. Environ. Dev.* **1998**, *7*, 45–59. [CrossRef]

28. Hadiz, V.R. Decentralization and democracy in Indonesia: A critique of neo-institutionalist perspectives. *Dev. Chang.* **2004**, *35*, 697–718. [CrossRef]

29. Resosudarmo, I.A.P. Closer to people and trees: Will decentralization work. In *Democratic Decentralization through a Natural Resource Lens*; Routledge: London, UK, 2005; pp. 110–132.

30. Cernea, M.M.; Schmidt-Soltau, K. The end of forcible displacements? Conservation must not impoverish people. *Policy Matters* **2003**, *12*, 42–51.

31. Sayer, J.; Bull, G.; Elliott, C. Mediating forest transitions: 'Grand design' or 'Muddling through'. *Conserv. Soc.* **2008**, *6*, 320. [CrossRef]

32. Kakabadse, Y. Involving communities: The role of NGOS. In *The Future of IUCN: The World Conservation Union*; IUCN: Gland, Switzerland, 1993; pp. 79–83.

33. McNeely, J.A. IUCN and indigenous peoples: How to promote sustainable development. In *The Cultural Dimension of Development: Indigenous Knowledge Systems*; Warren, D.M., Slikkerveer, L.J., Brokensha, D., Eds.; Intermediate Technology Publications Ltd: Rugby, UK, 1995; pp. 445–450.

34. WWF. *Conservation with People*; WWF: Gland, Switzerland, 1993.

35. Drew, J.A. Use of traditional ecological knowledge in marine conservation. *Conserv. Biol.* **2005**, *19*, 1286–1293. [CrossRef]

36. Boedhihartono, A.K. Dilemme à Malinau, Bornéo: Être ou ne pas être un chasseur-cueilleur Punan. Ph.D. Thesis, Université Paris VII, Paris, France, July 2004.

37. Levang, P.; Dounias, E.; Sitorus, S. Out of the forest, out of poverty? *For. Trees Livelihoods* **2005**, *15*, 211–235. [CrossRef]

38. Levang, P.; Sitorus, S.; Dounias, E. City life in the midst of the forest: A Punan hunter-gatherer's vision of conservation and development. *Ecol. Soc.* **2007**, *12*, 18. [CrossRef]

39. Gadgil, M.; Berkes, F.; Folke, C. Indigenous knowledge for biodiversity conservation. *Ambio* **1993**, *22*, 151–156.

40. Colding, J.; Folke, C. The relations among threatened species, their protection, and taboos. *Conserv. Ecol.* **1997**, *1*, 6. [CrossRef]

41. Johannes, R.E. The case for data-less marine resource management: Examples from tropical nearshore finfisheries. *Trends Ecol. Evolut.* **1998**, *13*, 243–246. [CrossRef]
42. Alcorn, J.B. Process as resource. *Adv. Econ. Bot.* **1989**, *7*, 1–63.
43. Berkes, F. *Sacred Ecology: Traditional Ecological Knowledge and Resource Management*; Taylor & Francis: Abingdon, UK, 1999.
44. Gadgil, M.; Rao, P.R.S.; Utkarsh, G.; Pramod, P.; Chhatre, A. New meanings for old knowledge: The people's biodiversity registers program. *Ecol. Appl.* **2000**, *10*, 1307–1317. [CrossRef]
45. Huntington, H.P. Using traditional ecological knowledge in science: Methods and applications. *Ecol. Appl.* **2000**, *10*, 1270–1274. [CrossRef]
46. Boedhihartono, A.K.; Gunarso, P.; Levang, P.; Sayer, J. The principles of conservation and development: Do they apply in Malinau? *Ecol. Soc.* **2007**, *12*, 2. [CrossRef]
47. Wadley, R.L.; Colfer, C.J.P. Sacred forest, hunting, and conservation in West Kalimantan, Indonesia. *Hum. Ecol.* **2004**, *32*, 313–338. [CrossRef]
48. Brosius, J.P. Endangered forest, endangered people: Environmentalist representations of indigenous knowledge. *Hum. Ecol.* **1997**, *25*, 47–69. [CrossRef]
49. Dahl, A.L. Traditional environmental knowledge and resource management in new Caledonia. In *Traditional Ecological Knowledge: A Collection of Essays*; IUCN: Gland, Switzerland, 1989; pp. 57–66.
50. Edgerton, R. *Sick Societies: Challenging the Myth of Primitive Harmony*; The Free Press: New York, NY, USA, 1992.
51. Rea, A.M. Resource utilization and food taboos of sonoran desert peoples. *J. Ethnobiol.* **1981**, *1*, 69–83.
52. Alvard, M.S. Testing the "ecologically noble savage" hypothesis: Interspecific prey choice by piro hunters of amazonia peru. *Hum. Ecol.* **1993**, *21*, 355–387. [CrossRef]
53. Berkes, F.; Folke, C.; Gadgil, M. Traditional ecological knowledge, biodiversity, resilience and sustainability. In *Biodiversity Conservation*; Kluwer Academic Publishers: Dordrecht, The Netherlands, 1995; pp. 281–299.
54. Ichwandi, I.; Shinohara, T. Indigenous practices for use of and managing tropical natural resources: A case study on Baduy community in Banten, Indonesia. *Tropics* **2007**, *16*, 87–102. [CrossRef]
55. Davis, W.; Henley, T. *Penan: Voice for the Borneo Rainforest*; Western Canada Wilderness: Toronto, ON, Canada, 1990.
56. Dennis, R.A.; Colfer, C.J.P.; Puntodewo, A. Forest Cover Change Analysis as Proxy: Sustainability Assessment using Remote Sensing and GIS in West Kalimantan, Indonesia. In *People Managing Forests: The Links Between Human Well Being and Sustainability*; Colfer, C.J.P., Byron, Y., Eds.; Resources for the Future and CIFOR: Washington, DC, USA, 2001; pp. 362–387.
57. Yuliani, E.L.; Adnan, H.; Achdiawan, R.; Bakara, D.; Heri, V.; Sammy, J.; Salim, M.A.; Sunderland, T. The roles of traditional knowledge systems in orang-utan pongo spp. And forest conservation: A case study of Danau Sentarum, West Kalimantan, Indonesia. *Oryx* **2016**. [CrossRef]
58. Alcorn, J.B.; Bamba, J.; Masiun, S.; Natalia, I.; Royo, A.G. Keeping ecological resilience afloat in cross-scale turbulence: An indigenous social movement navigates change in Indonesia. In *Navigating Social-Ecological Systems: Building Resilience for Complexity and Change*; Cambridge University Press: Cambridge, UK, 2003; pp. 299–327.
59. Padoch, C.; Peluso, N.L. *Borneo in Transition: People, Forests, Conservation, and Development*; Oxford University Press: Oxford, UK, 1996.
60. Quinten, M.; Stirling, F.; Schwarze, S.; Dinata, Y.; Hodges, K. Knowledge, attitudes and practices of local people on Siberut Island (West-Sumatra, Indonesia) towards primate hunting and conservation. *J. Threat. Taxa* **2014**, *6*, 6389–6398. [CrossRef]
61. WWF. *Saving Siberut: A Conservation Master Plan*; WWF Indonesia Programme: Bogor, Indonesia, 1980.
62. Whittaker, D.J. A conservation action plan for the mentawai primates. *Primate Conserv.* **2006**. [CrossRef]
63. Delfi, M. Islam and Arat Sabulungan in Mentawai. *Al Jamiah J. Islam. Stud. Yogyak. Indones.* **2013**, *51*, 475–499. [CrossRef]
64. Schefold, R. Religious Conceptions on Siberut, Mentawai. Available online: http://www.mentawai.org/miscellaneous-articles/bibliographic-resources-for-the-mentawai-islands/ (accessed on 19 January 2017).
65. Schefold, R. The domestication of culture: Nation-building and ethnic diversity in indonesia. *Bijdragen tot de Taal-, Land-en Volkenkunde* **1998**, *154*, 259–280. [CrossRef]

66. Bakker, L. Foreign images in Mentawai: Authenticity and the exotic. *Bijdragen tot de taal-, Land-en Volkenkunde* **2007**, *163*, 263–288. [CrossRef]

67. Koentjaraningrat. *Masyarakat terasing di Indonesia*; Seri Etnografi Indonesia 4: Gramedia, Jakarta, 1993.

68. Tyson, A.D. Still striving for modesty: Land, spirits, and rubber production in Kajang, Indonesia. *Asia Pac. J. Anthropol.* **2009**, *10*, 200–215. [CrossRef]

69. Rössler, M. Striving for modesty: Fundamentals of the religion and social organization of the makassarese patuntung. *Bijdragen tot de Taal-, Land- en Volkenkunde* **1990**, *146*, 289–324. [CrossRef]

70. Sellato, B. The nomads of Borneo: Hoffman and "devolution.". *Borneo Res. Bull.* **1988**, *20*, 106–120.

71. Cunliffe, R.N.; Lynam, T.J.P.; Sheil, D.; Wan, M.; Salim, A.; Basuki, I.; Priyadi, H. Developing a predictive understanding of landscape importance to the Punan-Pelancau of East Kalimantan, Borneo. *Ambio* **2007**, *36*, 593–599. [CrossRef]

72. Sheil, D.; Puri, R.K.; Basuki, I.; van Heist, M.; Wan, M.; Liswanti, N.; Sardjono, M.A.; Samsoedin, I.; Sidiyasa, K.; Permana, E. *Exploring Biological Diversity, Environment, and Local People's Perspectives in Forest Landscapes: Methods for a Multidisciplinary Landscape Assessment*; CIFOR: Kota Bogor, Indonesia, 2002.

73. Sheil, D.; Puri, R.; Wan, M.; Basuki, I.; Heist, M.v.; Liswanti, N.; Rachmatika, I.; Samsoedin, I. Recognizing local people's priorities for tropical forest biodiversity. *AMBIO J. Hum. Environ.* **2006**, *35*, 17–24. [CrossRef]

74. Dounias, E.; Froment, A. When forest-based hunter-gatherers become sedentary: Consequences for diet and health. *Unasylva* **2006**, *57*, 26–33.

75. Wunder, S.; Campbell, B.; Frost, P.G.H.; Sayer, J.A.; Iwan, R.; Wollenberg, L. When donors get cold feet: The community conservation concession in Setulang (Kalimantan, Indonesia) that never happened. *Ecol. Soc.* **2008**, *13*, 12. [CrossRef]

76. Sayer, J.A.; Margules, C.; Boedhihartono, A.K.; Sunderland, T.; Langston, J.D.; Reed, J.; Riggs, R.; Buck, L.E.; Campbell, B.M.; Kusters, K.; et al. Measuring the effectiveness of landscape approaches to conservation and development. *Sustain. Sci.* **2016**. [CrossRef]

land

MDPI

Article

Estate Crops More Attractive than Community Forests in West Kalimantan, Indonesia

James D. Langston [1],*, Rebecca A. Riggs [1],*, Yazid Sururi [1], Terry Sunderland [2] and Muhammad Munawir [3]

[1] College of Science & Engineering, James Cook University, Cairns, QLD 4870, Australia; yazid.sururi@my.jcu.edu.au
[2] Center for International Forestry Research (CIFOR), Bogor 16115, Indonesia; T.Sunderland@cgiar.org
[3] WWF-Indonesia, West Kalimantan Programme, Jalan Karna Sosial Gang Wonoyoso II No. 3, Pontianak 78121, Indonesia; nawir_1778@yahoo.co.id
* Correspondence: james.langston@my.jcu.edu.au (J.D.L.); rebecca.riggs@my.jcu.edu.au (R.A.R.); Tel.: +61-7-423-215-29 (J.D.L. & R.A.R.)

Academic Editors: Jeffrey Sayer and Chris Margules
Received: 19 January 2017; Accepted: 4 February 2017; Published: 8 February 2017

Abstract: Smallholder farmers and indigenous communities must cope with the opportunities and threats presented by rapidly spreading estate crops in the frontier of the agricultural market economy. Smallholder communities are subject to considerable speculation by outsiders, yet large-scale agriculture presents tradeoffs that they must navigate. We initiated a study in Sintang, West Kalimantan in 2012 and have returned annually for the last four years, building the baselines for a longer-term landscape approach to reconciling conservation and development tradeoffs in situ. Here, the stakeholders are heterogeneous, yet the land cover of the landscape is on a trajectory towards homogenous mono-cropping systems, primarily either palm oil or rubber. In one village on the frontier of the agricultural market economy, natural forests remain managed by the indigenous and local community but economics further intrude on forest use decisions. Conservation values are declining and the future of the forest is uncertain. As such, the community is ultimately attracted to more economically attractive uses of the land for local development oil palm or rubber mono-crop farms. We identify poverty as a threat to community-managed conservation success in the face of economic pressures to convert forest to intensive agriculture. We provide evidence that lucrative alternatives will challenge community-managed forests when prosperity seems achievable. To alleviate this trend, we identify formalized traditional management and landscape governance solutions to nurture a more sustainable landscape transition.

Keywords: conservation development tradeoffs; smallholder agriculture; agricultural market frontiers; community-based forestry; landscape approach

1. Landscapes in the Heart of Borneo

Communities in the Heart of Borneo (HoB), West Kalimantan, Indonesia are receiving international attention from the work of activist groups and action-research scientists [1]. External discourse often deafens us to the articulated perceptions of local people's lives and landscapes [2,3]. Mostly, the discourse victimizes people and their landscapes, subjecting them to scrutiny over their socioeconomic disadvantages. Science often situates their problems at the frontier of agribusiness economies and Indonesia's problematic, and often complex, governance arrangements [4–7]. A common concern entry-point emerges from the challenges and opportunities of new and rapidly-expanding oil palm plantations [8]. Locally, the millions of people living there ultimately face the consequences of this change [9]. Nationally, Indonesia generally prioritizes economic growth over achieving conservation goals [10]. Internationally, the environmental community focuses on enhancing and maintaining global public goods [11,12]. Communities in the HoB inhabit some of Indonesia's most dynamic frontier land. They are often Indonesia's poorest people, reside in the world's largest transboundary rainforest and face some of the world's greatest rates of deforestation due to the rapid expansion of oil palm [13–15].

The HoB is a tri-national transboundary initiative led by the World Wide Fund for Nature (WWF). Their coordinated efforts aim to sustainably manage landscapes for increased prosperity and biodiversity in Borneo's geographical center [8]. The HoB initiative aims to coordinate Indonesia, Brunei Darussalam and Malaysia to achieve the long term goal of conserving one of the world's most regarded biodiversity hotspots [16]. The initiative was established in response to high rates of forest conversion and degradation. The extent of Borneo's forests declined by 34% from 1973–2015, primarily due to agricultural expansion and El Niño Southern Oscillation (ENSO)-induced wildfires [17,18]. By 2015, Kalimantan (Indonesian Borneo) contained 5.7 million hectares of industrial plantations [18]. Oil palm drives the majority of agricultural expansion in Kalimantan [19]. In 2010, in West Kalimantan, more than half a million hectares of oil palm were under cultivation, with a planned 5 million more hectares, already under concession [1]. Then, 79% of allocated oil palm leases remained undeveloped [10]. Projections show that full development would convert approximately 90% of remaining available forest lands [20]. Oil palm would occupy 34% of lowlands outside protected areas (ibid). Realistically, the development of the oil palm sector receives greater governmental support than meeting conservation targets, including those underlining the HoB initiative [10]. In 2011, the oil palm industry contributed US $20 billion in foreign exchange earnings to Indonesia [14].

To increase conservation impact, HoB operational management is devolved to the landscape scale—a spatial delineation defined by combination of social-ecological parameters including watersheds and political jurisdictions [8]. This stems from evidence that biodiversity and environmental conservation action also aimed at addressing the aspirations and poverty of locals is best addressed at the landscape scale [21]. However, while conservationists lament the rapidly increasing pressures on tropical landscapes from increasing global agricultural needs, agricultural investment is often the only opportunity available to meet rising development aspirations of rural forest dwellers [9,22,23]. Agricultural innovation at the landscape scale must benefit smallholders for inclusive development, a pre-requisite to achieving long-term conservation goals if they are to remain living there [24,25]. In West Kalimantan, local management of resources includes both indigenous management and community-based management. In these local landscape contexts, customary forests have new found legal support for local decision-making and user rights. Conservation and development organizations will need to come to terms with the choices that these groups make for their own interests. The problem is evident: assumptions and expectations of development and conservation are clearly at odds in this landscape.

We have worked with WWFs regional Sintang and Kapuas Hulu offices in West Kalimantan as an entry point for building landscape approach platform for action-research. Ownership and power are hotly contested issues within this landscape [26] and thus taking a landscape approach provides a framework to make progress toward achieving satisfactory outcomes for the broad range of stakeholders concerned [27]. We hypothesize that conservation efforts will fail if local people remain living in poverty. We ask: *what will happen to forests in the control of local people when development opportunities arise?* We use the case study of Kenyabur Baru village in Sintang Regency, West Kalimantan to demonstrate that community forest management fails when the economic returns of converting forest to oil palm exceed those of intact forest. The following reports on lessons from our observations in the Sintang Regency as part of the HoB initiative.

2. Conceptual Framework

The landscape approach is the latest iteration of attempts to integrate conservation and development in defined geographic spaces [28]. A landscape approach is defined as "*A long-term collaborative process bringing together diverse stakeholders aiming to achieve a balance between multiple and sometimes conflicting objectives in a landscape or seascape"* [28]. The landscape approach seeks to address global challenges of poverty alleviation, food security, climate change, and biodiversity loss [29]. Though it is a refinement of prior approaches, it is distinct as it explicitly acknowledges that satisfying all stakeholders will often be unachievable. However, its aim is to manage these tradeoffs transparently through governance principles that aspire to reach consensus, whereas other approaches portend spurious 'win–win' outcomes by failing to acknowledge the magnitude of stakeholder diversity and the need for compromise and negotiation [27]. Primarily, landscape approaches are a question of governance. The most recent research identifies how landscape practitioners might measure governance processes, recognizing that process is vital to contextualizing and then achieving desirable and sustainable outcomes [28]. Landscape approaches provide a conceptual framework to make long-term improvements to conservation and production by engaging and empowering local stakeholders [30]. Capacity building, local empowerment, improving governance and providing transparency in resource management negotiations are fundamental components of landscape approaches [31–33].

2.1. Community Management in Indonesia's Landscapes

> *If we truly want to address issues of climate change, poverty, forest and biodiversity loss effectively, the global community will have to devote far greater efforts than has occurred to date to accessing the views, preferences, and goals of marginalized peoples, understanding local social systems, and incorporating such information into policies, laws, and regulations [2].*

The communities in the HoB are diverse. Smallholder oil palm communities in West Kalimantan are similarly heterogeneous. Likewise, indigenous groups, non-indigenous farming groups, and transmigrants live side-by-side. There are no simple typologies of oil palm farmers—yet in the discourse, generalizations abound [1,10,14,23]. There are also wide ranging opinions on the promise of community-based natural resource management to deliver environmental benefits [34]. However, policy fails if it is too top down and if it does not acknowledge and involve the power and interests of local people [35]. These communities make decisions over the use of their lands and their decision-making is in the context of rapidly spreading oil palm, stemming from large and intermediate-sized companies [36]. Large and mid-size companies provide economies of scale for smallholder participation in a cash crop economy, triggering its expansion (ibid).

There is a long convoluted history of land-use decision making in Indonesia [34]. Indigenous groups, currently through the National Alliance of Indigenous Peoples (Aliansi Masyarakat Adat Nusantara, AMAN), recently succeeded in moving policy agendas beyond community forest management. They now call for an end to State control over customary land. AMAN defines community indigenous peoples (*masyarakat adat*) as "communities living on the basis of ancestral origins in an adat region, that have sovereignty over land and natural resource wealth; a sociocultural life regulated by adat law; and an adat council that manages the daily life of its people" [37]. By 2014, AMAN claimed to be representing well over 2,000 indigenous communities [37]. Laws to remove customary forests from state control were codified by the Indonesian Constitutional Court in May 2013. Muddying the waters, a powerful union of peasant groups also claim development-related rights and responsibilities over resources. AMAN and the peasant unions campaigning for agrarian reform have very different understandings about claims to adat lands [37]. AMAN wants to reclaim land for 'indigenous' groups who have, by their definition, historic and collective rights to it. The unions however aim to recover as much land as possible to redistribute for poorer communities—indigenous, otherwise local, and migrant alike. The federal land allocation agency codified the ambiguity over adat claims to land when the Ministry of Environment and Forestry issued Ministerial Regulation No. 9 of 2015, that simplifies the concept of indigenous rights into communal rights [38]. Although securing land for the poor and marginalized is noble, the process encourages more groups to claim adat land and to manage it as they wish, within a very broad range of contexts. The implementation of customary laws in ambiguous contexts causes concern among those who worry about the future of environmental assets, reviving an old fear of the 'tragedy of the commons' [34].

2.2. Flawed Assumptions

Confounding the issue, it is abundantly clear that hierarchies of power do not share the same realities. According to Astuti and McGregor [39], a federal-level stakeholder management leader stated that 'indigenous people owned the wisdom of treating the forest with care, the wisdom that respects nature and the cultural spiritual values'. Indigenous knowledge and wisdom is revered for living harmoniously with nature. This perception of 'the indigenous' reality is a spurious caricature disconnected to reality; indigenous groups also want to benefit from extractive industries and modernity [40]. The assumption that indigenous people are bound to be 'green' has led to conservation organizations associated with concerns about green grabbing strategically engaging with indigenous activist organizations to pursue land claims [39]. Our data shows that in West Kalimantan, heterogeneous communities possess multiple interests, including benefitting from estate crop development at the expense of forest [39].

So-called 'green grabbing conservation organizations' (green grab being a style of land grab to ostensibly pursue conservation or environmental outcomes as core objectives) have also perpetuated the notion that conservation can succeed in the long run in places where people continue to live in poverty [41–43]. This is contrary to evidence that while a population is living in poverty, they will continue to exert pressure on natural resources with negative conservation outcomes [44–46]. More egregiously, conservation efforts can inhibit development pathways and fail in areas where poverty persists [47–50]. We examine these interactions in our study and hypothesize that conservation efforts will fail if local people remain living in poverty.

3. Methods

In 2012 we began applying landscape approach principles [27] to engage and assess landscape level interventions in the Sintang Regency. WWF was our institutional entry-point for building landscape-level governance coalitions, with whom we had previous collaboration. We assert that building a landscape-level process of determining objectives, measuring progress to meet those objectives, and reflecting on lessons learned must be undertaken with the participation of all stakeholders [28]. The sustainable livelihoods framework's capital assets provide our framework for determining landscape explicit assets [51]. We conducted participatory modeling to begin to allow for scientific rigor in establishing the links between interventions and outcomes [52–54]. The process was driven by a multi-stakeholder forum. The forum comprised of representatives of conservation and development organizations (local non-governmental organizations (NGOs), international research organizations and private industry), staff of landscape level government agencies (sub-district and village level), and local people from communities where our WWF connections allowed access. Gender was accounted for in settings both through mixed gender and isolated gender focus group discussions and interviews. In 2015 we provided training in simple modeling techniques using the software STELLA [53]. At this initial meeting, we decided that long-term perception data in villages would prove useful. We therefore utilize villages as sentinel sites for setting up long term panel data [55]. Our panel data is based on interviews with local key informants over the last three years. We interviewed respondents using an interpretivist approach, using a general inductive method [56]. By doing so, we sought answers to specific questions but exercised considerable flexibility to enable exploration of unanticipated issues that may arise. We also held focus group discussions (FGD) around topics of interest, maintaining the same approach as our semi-structured interviews. The working languages of our group were English and Indonesian. The data used for this paper is based on interviews within one village at one end of a landscape transition, they retained forests over which they exercise their adat rights. This paper is based on recent visits and the fledgling panel data to a series of villages, focusing on one village, Kenyabur Baru in the Sintang district, where community-based forest management remains part of their social-ecological system.

4. Results

4.1. Sintang Case: Development Opportunities Arise

In Kenyabur Baru and nearby villages, transmigrants, indigenous, and local farmers live side by side. Local farmers identified themselves as locally indigenous, i.e. of the local Dayak clan group, or from other clan groups that were not historically from that land, whereas the term "transmigrant" refers to the government-led re-location schemes. Spontaneous and government-sponsored transmigration has brought waves of Javanese migrants to the area since the early 1900s. The indigenous Dayak populations currently co-exist with migrants and there is a diverse ethnic mix in the area. Dayaks have adopted Javanese cultural ways whilst migrants adopt those of the local Dayaks. Oil palm and rubber are the dominant agricultural endeavors but livelihood outcomes are not homogenous; each family engages in a unique way. This does not fit into the neat dual business model typology of either 'tethered scheme' (plasma) smallholders or independent smallholders. Plasma smallholders usually receive credit from a plantation for planting and inputs. Independent smallholders are unassisted but are dependent on an estate mill to process their fruit. Locals have land in estates through various terms of engagement, and possess land locally. Local respondents indicate that government rules matter much less than local arrangements in the community and between companies with which they have profit sharing/crop agreements. Local people cut forests for either larger companies or themselves, and their cropping size ranges from two to 50 hectares. Land assets do not reflect ethnicity but instead local power relations, which are ethnically mixed; those possessing the greatest social capital in the area are the most land and resource rich. These richer farmers are early adapters of agricultural innovations and are most connected with the town of Sintang and external markets.

> "I have connections in the city and think that the opportunity to live well out here depends on my willingness to be opportunistic, investing in expansion and experimentation so that risk is counterbalanced by delivering products to market, and my freedom to choose that market." (panel data respondent No. 1)

Until oil palm arrived in the 1990s, most of these villages practiced swidden rice cultivation and had plots of rubber agroforests. Consequently, much of the land now occupied by oil palm plantations in this landscape had previously been managed as rubber agroforests. There was very little old growth forest (Figures 1 and 2). Locals have increasingly abandoned traditional shifting agriculture due to increased land pressure from rising population and establishments of estate crops by larger companies.

> "There is not enough land left for us to do what we use to do, some of that is because of our population expansion, some of that is because companies now own large plantations." (FGD respondent)

If local communities can accumulate more land, they prioritize rubber or mono-cropping of oil palm. More recent migrants from Java are more likely to plant something new, such as oil palm in an otherwise rubber-dominated landscape, as they have connection to companies and have been exposed to contemporary industrial processes and economies. A co-operative based in one village with a Javanese leader will often try to push greater oil palm engagement, but many local people still prefer to farm rubber, a practice with years of accumulated knowledge. More migrants have arrived recently, as the promise of prosperity from cash-crops and growing social-networks provided socioeconomic pulls to the forest frontier.

Figure 1. Mosaic of land uses including oil palm and rubber surrounding Kenyabur Baru.

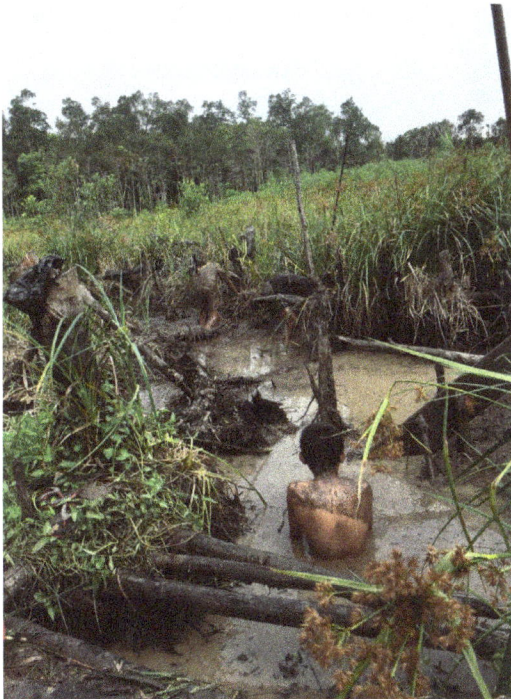

Figure 2. Children clearing root vegetables from newly cleared peatland drainage ditch. The area is being prepared for a rubber plantation.

4.2. A Forest at Stake

We visited Kenyabur Baru to observe the adat forest (Figure 3). Kenyabur Baru is a frontier village, specifically an agricultural market frontier village [42]. Frontier and disputed areas are where pressures for deforestation and degradation are increasing, and control is often insecure and in conflict. Many of the villages closer to the district capital city Sintang have no more natural forests of significant size left. The regional villages are increasingly tied to the economy of Sintang. Sintang's economic growth is primarily linked to the growth of industrial oil palm plantations; rural communities are increasingly participating in this economy. Roads have been developed by the government to get products to market and to access services. The roads are in poor condition—they are unpaved and only accessible by four-wheel drive vehicles or motorbikes. The community desires better access to markets and services via improved roads and economic networks.

> "The road has existed for a long time, without it we would not be here doing what we do now. But we want more, we want to be able to reach markets, we want paved roads so that we are safe in cases of emergencies and for easier day to day lifestyle." (panel data respondent No. 2)

Figure 3. The village maintains adat-managed forests, which lie adjacent to the end of their road. The beginning of the forest can be seen in the left lower hand of the image.

The adat forest in Kenyabur Baru retains high conservation value. Of high conservation value are *Shorea seminis Slooten* (critically endangered) and the *Shorea stenoptera Burck* (endangered). High social and cultural conservation values exist in the forest as food resources, traditional medicine and home-building materials. The adat forest is an old growth, minimally-used forest. There is evidence of large mammals. We observed sun bear markings on trees and locals report recent and regular but diminishing sightings of pig-tailed macaques (Figure 4). However, elders last observed orangutans in their forests more than a generation ago. There are mature strangling figs, abundant lianas and other secondary regrowth due to previous forest clearance, and diverse and abundant mature dipterocarps. The forest sits on peatlands approximately 1m deep, has a leaf litter depth of 15–25 cm, and has a mature complex structure.

Figure 4. Recent sun bear markings on a tree inside the adat forest.

Focus group discussions informed us that the forest adjacent to the community land remains adat forest due to the cultural values they derive from it. While the focus group discussion was comprised of indigenous and migrants, both recent and old, they affirmed a mutual communal attachment to the forest. This attachment is based in benefits provided to them. Benefits provided include non-timber forest products (NTFPs), ecosystem services, and rarely, timber for cash. The NTFPs do not generate income but are used for ceremonial or medicinal purposes (Figure 5). It is prohibited to cut down trees except during financial emergencies when people can sell felled trees to pay for health or schooling (they cited a case of a health emergency). They acknowledged, without prompting, that the forest also benefits them through other provisional services such as micro-climate benefits and watershed stability.

Figure 5. Adat elder showing ceremonial plants found in the adat forest.

For the time being, local adat culture impels those of Kenyabur Baru to maintain forest even if other land uses seem more lucrative to them. However, economics is further intruding on their decision to manage the forests for adat value. In our focus group discussion, there was consensus that the value of converting adat forests to either palm oil or rubber plantation exceeded the value of forests as they stand. When asked why they had not cleared more, they claimed they were waiting until improved seeds became affordable and accessible. Villagers also claimed it would be too arduous to clear the land but they welcomed help to clear it (they did not identify burning the forest as a potential and easy clearing method). Priorities are not the same now as they were in the past; values have changed with proximity to the agricultural market economy. As roads and associated spillover infrastructure have developed, the village has become more integrated into market economies wherein the benefits of engaging with the market economy are more apparent. According to villagers, children suddenly had opportunities to go to schools, healthcare was better, and information technology put the visions and accessibility of modern amenities within reach.

> "The economic opportunities provided by road access originally stimulated by the oil palm industry has made life better. There are some social costs but we all now have a desire for modern amenities and want to live prosperously." (panel data respondent No. 3)

Villagers acknowledge that accessing these amenities means greater participation in the cash economy and that this is incompatible with more traditional livelihood activities. During a ranking exercise, the community prioritized rubber above oil palm as a preferred land use—it was the highest priority land use option for them. As stated earlier, rubber provides daily income, something more valuable than less frequent value chain payoff commodities such as oil palm. They also identified freedom and independence over their silvicultural practices and choice of buyers and middle-men as major reasons for preferring rubber. However, the villagers also contextualized their preferences for rubber. In the present situation, they lack capital, labor and power to manage oil palm. They foresee that with greater incomes, greater connectivity to market with better roads, and with social capital remaining strong, they will convert existing rubber to oil palm. The heterogeneity of the community and their relative 'development' isolation has not led to simple patterns of adat vs local vs transmigrant values in the landscape. Rather, similarities emerged: they firstly aspire for capital reliability. Secondly, once they have reliable incomes and safety nets, they aspire for capital accumulation. Thirdly, they aspire to capital re-investment for their kin.

However, younger community members have a different vision for their future. They foresee a landscape void of smallholders and villagers. The alluring amenities they can see on the internet do not seem as out of reach as they do to their elders. Elders describe a future wherein their progeny can have better access to education and can live better lives without abandoning social values based in adat culture. Young people increasingly regard urbanization processes as desirable.

> "We would ideally choose office jobs but invest in land. We want some forests to remain, but want to profit from our lands and while living in the city. In 100 years there will be no people living here anymore. They will all be either working on plantations or in jobs in the city." (panel data respondent No. 4)

5. Conclusions

Local people almost always express a strong desire for development and lament their few opportunities .... [1]

While locally managing the forest in Kenyabur Baru has succeeded in maintaining biodiversity and conservation values, maintaining adat management now appears less attractive to the local community than conversion to estate crops—rubber or oil palm. Many other poor rural communities within the HoB find that managing forests is less profitable than intensive agriculture, and the communities desire prosperity and development [1]. This case illustrates how poverty is a threat to community-managed conservation success when profitable opportunities to convert forest lands present themselves. While communities aspire to conserve environmental and cultural values, this desire is outweighed by economic factors. The communities' willingness to court more economically attractive uses for the adat land for either oil palm or rubber illustrates how community-based resource management can fail. This is in line with the arguments that biodiversity provides few instrumental values for poor people [57]. As the village furthest from the city with poor road access, those of Kenyabur Baru do indeed lament their few opportunities to develop and they are not going to keep the forest if they can derive benefits from other land uses. Currently there are no other mechanisms offering an equivalent pathway to livelihood improvements.

In the HoB, as agricultural markets approach frontier landscapes, forest dwellers and smallholder communities have transitioned from shifting cultivation and timber production to more sedentary extraction and intensified agriculture. Here, the realities of new economic frontiers force community forestry management to adapt to increasing pressures or they will not succeed. Conservation organizations need to recognize the extent of these tradeoffs and the mode by which local people determine land use if they wish to engage in community driven conservation. Similarly, if community advocacy groups in Indonesia fail to acknowledge local heterogeneity, the desire for development and agency with which local communities determine land use, collaboration or collusion between them will be weak, and outcomes will be unsatisfactory. Past successes of adat management in maintaining the biodiversity in forests are unlikely to be replicable in the face of lucrative alternatives. Adat management cannot be kept separate from modern incentive systems. If elders and community members can obtain benefits from adat management in the face of economic pressures, there could be room for innovation in the form of adat formalization.

An example of successful adat formalization where conservation values were retained is in Danau Empangau, Kapuas Hulu, West Kalimantan [58]. By garnering support from district heads, the community succeeded in checking the power of the industrial actors, reconciling power asymmetries in the landscape. Adat management of high conservation value resources succeeded by coexisting with industrial corporate estate cropping in a spatially optimal way. Other examples exist where the evolution of adat power is decentralized and empowered at a community level and can coexist with formal resource management systems within formal structures of land use governance [59–61]. We assert that governance processes in the form of a landscape approach must be applied to enable this process because it provides a framework and guidance on good practice for landscape processes. Multi-criteria assessments can provide tools for achieving spatially optimal solutions that empower adat management through a landscape approach process [62,63]. Landscape scale governance learning processes could adhere to new measurement principles that ensure societal beneficial landscape outcomes [28].

Poverty and deforestation historically have shared a win–lose relationship, meaning that deforestation is the price of development [46]. The win–lose trajectory has historically been associated with rural development: the conversion of forest to intensive agriculture. In that scenario forests shrink but employment and incomes increase [64]. To reach landscape transition, forest cover must rebound without having lost its ecological memory. This would best approximate a win–win outcome in the long run. Forest conversion resulting in unprofitable agriculture, only providing subsistence or ephemeral income to a poor population that might be even worse off if they were cut off from the market economy, would be the worst, lose–lose case for forests and people.

Landscape management coalitions should direct their effort on improving livelihoods, moving through a forest transition, and maintaining conservation values within a mosaic of different lands uses. Multi-stakeholder forums as part of a management coalition in a landscape approach must engage in good process management to reconcile the local socioeconomic pressures with external drivers of change. Good process management includes negotiation and communication of clear goals, a clear and agreed theory of change, a rigorous and equitable process for continuing stakeholder engagement, connection to policy processes and key actors, effectiveness of governance, and transparency [28].

These should be measured for continual landscape learning, for better evidence-based decision-making. Adat management can coexist within more formal land tenure arrangements, but must benefit from a legal recognition of its role in current formal governance structures. It is clear that organizations must recognise that under some conditions the benefits of deforestation and infrastructure development may outweigh the costs [65]. The governance of local forest resources for long-term gain is only possible if local stakeholders are committed to conservation goals and these goals are supported by durable policy arrangements. If local community and indigenous groups are stakeholders with longer-term commitments for stewardship of resources than politicians making unpredictable volatile policy environments, then they need to be a major driver of policy durability. In Indonesia this is problematic because of complexities and contestations over local lands. AMAN groups have championed the rights of indigenous communal land ownership, yet a competing 'community rights' organization has also championed rights of local communities. A landscape approach wherein a management coalition coordinates visions between actors and agents in the landscape will provide a backbone for durable policymaking.

Presently, there are insufficient institutions in place to guarantee that the forests are managed sustainably. Governance must include coordinated visions and address development needs but forested lands will suffer losses in the face of more profitable endeavors if they are wholly managed by communities in an agricultural market frontier. Ideological arguments that have dominated the discourse and have polarized the conservation and community rights advocates must be met with evidence. Our evidence shows that conservation will not succeed in a community that wants the benefits of more financial prosperity when development opportunities arise.

Acknowledgments: We would like to thank all our colleagues in the HoB. We especially thank CIFOR for funding and other in-kind assistance. We are grateful to Albertus Albertus (WWF), Aseop Asep Bee (WWF), and all our friends in Kenyabur Baru, Mererai 1, and Mererai 2. We also thank the two reviewers for their valuable comments on this manuscript.

Author Contributions: James D. Langston conceived and designed the study; James D. Langston and Yazid Sururi performed the field study; Muhammad Munawir helped facilitate field study and contributed to the biophysical and social data; James D. Langston, Rebecca A. Riggs and Yazid Sururi analyzed the data; Terry Sunderland provided insightful edits and ideas to enrich the context, results, and discussion. James D. Langston and Rebecca A. Riggs drafted the original paper and Yazid Sururi, Muhammad Munawir, Terry Sunderland contributed edits and comments to subsequent drafts.

Conflicts of Interest: The authors declare no conflict of interest.

References

1. Levang, P.; Riva, W.F.; Orth, M.G. Oil palm plantations and conflict in Indonesia: Evidence from West Kalimantan. In *The Oil Palm Complex: Smallholders, Agribusiness and the State in Indonesia and Malaysia*; NUS Press: Singapore, 2016; pp. 283–300.
2. Colfer, C.J.P. Marginalized forest peoples' perceptions of the legitimacy of governance: An exploration. *World Dev.* **2011**, *39*, 2147–2164. [CrossRef]
3. Meijaard, E.; Abram, N.K.; Wells, J.A.; Pellier, A.-S.; Ancrenaz, M.; Gaveau, D.L.; Runting, R.K.; Mengersen, K. People's perceptions about the importance of forests on Borneo. *PLoS ONE* **2013**, *8*, e73008. [CrossRef] [PubMed]
4. Peluso, N.L. The political ecology of extraction and extractive reserves in East Kalimantan, Indonesia. *Dev. Change* **1992**, *23*, 49–74. [CrossRef]
5. Bullinger, C.; Haug, M. In and out of the forest: Decentralisation and recentralisation of forest governance in East Kalimantan, Indonesia. *Austrian J. South-East Asian Stud.* **2012**, *5*, 243–262.
6. Gallemore, C.T.; Rut Dini Prasti, H.; Moeliono, M. Discursive barriers and cross-scale forest governance in Central Kalimantan, Indonesia. *Ecol. Soc.* **2014**. [CrossRef]
7. Myers, R.; Sanders, A.J.; Larson, A.M.; Ravikumar, A. *Analyzing Multilevel Governance in Indonesia: Lessons for REDD+ from the Study of Landuse Change in Central and West Kalimantan*; CIFOR (Center for International Forestry Research): Bogor, Indonesia, 2016.

8. Hitchner, S.L. Heart of Borneo as a 'Jalan Tikus': Exploring the links between indigenous rights, extractive and exploitative industries, and conservation at the world conservation congress 2008. *Conserv. Soc.* **2010**, *8*, 320–330. [CrossRef]

9. Sayer, J.; Cassman, K.G. Agricultural innovation to protect the environment. *Proc. Natl. Acad. Sci. USA* **2013**, *110*, 8345–8348. [CrossRef] [PubMed]

10. Potter, L. *Managing Oil Palm Landscapes: A Seven-Country Seurvey of the Modern Palm Oil Industry in Southeast Asia, Latin America and West Africa*; CIFOR (Center for International Forestry Research): Bogor, Indonesia, 2015.

11. Carlson, K.M.; Curran, L.M.; Asner, G.P.; Pittman, A.M.; Trigg, S.N.; Adeney, J.M. Carbon emissions from forest conversion by Kalimantan oil palm plantations. *Nature Clim. Change* **2013**, *3*, 283–287. [CrossRef]

12. Boons, F.; Mendoza, A. Constructing sustainable palm oil: How actors define sustainability. *J. Clean. Prod.* **2010**, *18*, 1686–1695. [CrossRef]

13. Lindsay, E.; Convery, I.; Ramsey, A.; Simmons, E. Changing place: Palm oil and sense of place in Borneo. *Hum. Geogr.* **2012**, *6*, 45–53. [CrossRef]

14. Byerlee, D.; Naylor, R.L. *The Tropical Oil Crop Revolution: Food, Feed, Fuel, and Forests*; Oxford University Press: Oxford, UK, 2016.

15. Runting, R.K.; Meijaard, E.; Abram, N.K.; Wells, J.A.; Gaveau, D.L.A.; Ancrenaz, M.; Possingham, H.P.; Wich, S.A.; Ardiansyah, F.; Gumal, M.T.; et al. Alternative futures for borneo show the value of integrating economic and conservation targets across borders. *Nature Commun.* **2015**. [CrossRef] [PubMed]

16. WWF. Heart of Borneo. Available online: http://wwf.panda.org/what_we_do/where_we_work/borneo_forests/ (accessed on 15 November 2016).

17. Gaveau, D.L.; Sloan, S.; Molidena, E.; Yaen, H.; Sheil, D.; Abram, N.K.; Ancrenaz, M.; Nasi, R.; Quinones, M.; Wielaard, N. Four decades of forest persistence, clearance and logging on Borneo. *PLoS ONE* **2014**, *9*, e101654. [CrossRef] [PubMed]

18. Gaveau, D.L.; Sheil, D.; Husnayaen, M.A.S.; Arjasakusuma, S.; Ancrenaz, M.; Pacheco, P.; Meijaard, E. Rapid conversions and avoided deforestation: Examining four decades of industrial plantation expansion in Borneo. *Sci. Rep.* **2016**. [CrossRef] [PubMed]

19. Deakin, L.; Kshatriya, M.; Sunderland, T. *Agrarian Change in Tropical Landscapes*; CIFOR (Center for International Forestry Research): Bogor, Indonesia, 2016.

20. Carlson, K.M.; Curran, L.M.; Ratnasari, D.; Pittman, A.M.; Soares-Filho, B.S.; Asner, G.P.; Trigg, S.N.; Gaveau, D.A.; Lawrence, D.; Rodrigues, H.O. Committed carbon emissions, deforestation, and community land conversion from oil palm plantation expansion in West Kalimantan, Indonesia. *Proc. Natl. Acad. Sci. USA* **2012**, *109*, 7559–7564. [CrossRef] [PubMed]

21. Colfer, C.J.P.; Pfund, J.L. *Collaborative Governance of Tropical Landscapes*; Routledge: Abingdon, UK, 2011.

22. Laurance, W.F.; Sayer, J.; Cassman, K.G. Agricultural expansion and its impacts on tropical nature. *Trends Ecol. Evol.* **2014**, *29*, 107–116. [CrossRef] [PubMed]

23. Rist, L.; Feintrenie, L.; Levang, P. The livelihood impacts of oil palm: Smallholders in Indonesia. *Biodivers. Conserv.* **2010**, *19*, 1009–1024. [CrossRef]

24. Vadjunec, J.M.; Radel, C.; Turner II, B. Introduction: The continued importance of smallholders today. *Land* **2016**. [CrossRef]

25. Hettig, E.; Lay, J.; Sipangule, K. Drivers of households' land-use decisions-a critical review of micro-level studies in tropical regions. *Land* **2016**. [CrossRef]

26. Moeliono, M.; Limberg, G. *The Decentralization of Forest Governance: Politics, Economics and the Fight for Control of Forests in Indonesian Borneo*; Earthscan: London, UK, 2012.

27. Sayer, J.; Sunderland, T.; Ghazoul, J.; Pfund, J.L.; Sheil, D.; Meijaard, E.; Venter, M.; Boedhihartono, A.K.; Day, M.; Garcia, C. Ten principles for a landscape approach to reconciling agriculture, conservation, and other competing land uses. *Proc. Natl. Acad. Sci. USA* **2013**, *110*, 8349–8356. [CrossRef] [PubMed]

28. Sayer, J.A.; Margules, C.; Boedhihartono, A.K.; Sunderland, T.; Langston, J.D.; Reed, J.; Riggs, R.; Buck, L.E.; Campbell, B.M.; Kusters, K.; et al. Measuring the effectiveness of landscape approaches to conservation and development. *Sustain. Sci.* **2016**. [CrossRef]

29. Reed, J.; Van Vianen, J.; Deakin, E.L.; Barlow, J.; Sunderland, T. Integrated landscape approaches to managing social and environmental issues in the tropics: Learning from the past to guide the future. *Glob. Change Biol.* **2016**. [CrossRef] [PubMed]

30. Estrada-Carmona, N.; Hart, A.K.; DeClerck, F.A.; Harvey, C.A.; Milder, J.C. Integrated landscape management for agriculture, rural livelihoods, and ecosystem conservation: An assessment of experience from Latin America and the Caribbean. *Landsc. Urban Plan.* **2014**, *129*, 1–11. [CrossRef]

31. Smith, R.J.; Veríssimo, D.; Leader-Williams, N.; Cowling, R.M.; Knight, A.T. Let the locals lead. *Nature* **2009**, *462*, 280–281. [CrossRef] [PubMed]

32. Pfund, J.-L. Landscape-scale research for conservation and development in the tropics: Fighting persisting challenges. *Curr. Opin. Environ. Sustain.* **2010**, *2*, 117–126. [CrossRef]

33. Milder, J.C.; Hart, A.K.; Dobie, P.; Minai, J.; Zaleski, C. Integrated landscape initiatives for African agriculture, development, and conservation: A region-wide assessment. *World Dev.* **2014**, *54*, 68–80. [CrossRef]

34. Riggs, R.A.; Sayer, J.; Margules, C.; Boedhihartono, A.K.; Langston, J.D.; Sutanto, H. Forest tenure and conflict in Indonesia: Contested rights in Rempek Village, Lombok. *Land Use Policy* **2016**, *57*, 241–249. [CrossRef]

35. Bull, G.; Elliott, C.; Boedhihartono, A.; Sayer, J. Failures in tropical forest and conservation policy: What is the solution? *J. Trop. For. Sci.* **2014**, *26*, 1–4.

36. Li, T.M. *Social Impacts of Oil Palm in Indonesia: A Gendered Perspective from West Kalimantan*; CIFOR (Center for International Forestry Research): Bogor, Indonesia, 2015.

37. Institute for Policy Analysis of Conflict. *Indigenous Rights VS Agrarian Reform in Indonesia: A Case Study from Jambi*; Institute for Policy Analysis of Conflict: Jakarta, Indonesia, 2014.

38. Nusantara, A.M.A. *2015 Year-End Note*; University of Illinois: Champaign, IL, USA, 2015.

39. Astuti, R.; McGregor, A. Indigenous land claims or green grabs? Inclusions and exclusions within forest carbon politics in Indonesia. *J. Peasant Stud.* **2016**. [CrossRef]

40. Li, T.M. *Land's End: Capitalist Relations on an Indigenous Frontier*; Duke University Press: Durham, NC, USA, 2014.

41. Fairhead, J.; Leach, M.; Scoones, I. Green grabbing: A new appropriation of nature? *J. Peasant Stud.* **2012**, *39*, 237–261. [CrossRef]

42. McCarthy, J.F.; Vel, J.A.; Afiff, S. Trajectories of land acquisition and enclosure: Development schemes, virtual land grabs, and green acquisitions in Indonesia's outer islands. *J. Peasant Stud.* **2012**, *39*, 521–549. [CrossRef]

43. Corson, C.; MacDonald, K.I.; Neimark, B. Grabbing "green": Markets, environmental governance and the materialization of natural capital. *Hum. Geogr.* **2013**, *6*, 1–15.

44. Sandker, M.; Suwarno, A.; Campbell, B.M. Will forests remain in the face of oil palm expansion? Simulating change in Malinau, Indonesia. *Ecol. Soc.* **2007**, *12*, 37. [CrossRef]

45. Sandker, M.; Ruiz-Perez, M.; Campbell, B.M. Trade-offs between biodiversity conservation and economic development in five tropical forest landscapes. *Environ. Manag.* **2012**, *50*, 633–644. [CrossRef] [PubMed]

46. Sunderlin, W.D.; Angelsen, A.; Belcher, B.; Burgers, P.; Nasi, R.; Santoso, L.; Wunder, S. Livelihoods, forests, and conservation in developing countries: An overview. *World Dev.* **2005**, *33*, 1383–1402. [CrossRef]

47. Adams, W.M.; Aveling, R.; Brockington, D.; Dickson, B.; Elliott, J.; Hutton, J.; Roe, D.; Vira, B.; Wolmer, W. Biodiversity conservation and the eradication of poverty. *Science* **2004**, *306*, 1146–1149. [CrossRef] [PubMed]

48. Pimbert, M.P.; Ghimire, K. *Social Change and Conservation: Environmental Politics and Impacts of National Parks and Protected Areas*; Earthscan Publications: London, UK, 1997.

49. Colchester, M. *Salvaging Nature: Indigenous Peoples, Protected Areas and Biodiversity Conservation*; Diane Publishing: Collingdale, PA, USA, 1994.

50. Norton-Griffiths, M.; Southey, C. The opportunity costs of biodiversity conservation in Kenya. *Ecol. Econ.* **1995**, *12*, 125–139. [CrossRef]

51. Bebbington, A. Capitals and capabilities: A framework for analyzing peasant viability, rural livelihoods and poverty. *World Dev.* **1999**, *27*, 2021–2044. [CrossRef]

52. Sandker, M.; Campbell, B.M.; Nzooh, Z.; Sunderland, T.; Amougou, V.; Defo, L.; Sayer, J. Exploring the effectiveness of integrated conservation and development interventions in a central African forest landscape. *Biodivers. Conserv.* **2009**, *18*, 2875–2892. [CrossRef]

53. Sandker, M.; Campbell, B.M.; Ruiz-Pérez, M.; Sayer, J.A.; Cowling, R.; Kassa, H.; Knight, A.T. The role of participatory modeling in landscape approaches to reconcile conservation and development. *Ecol. Soc.* **2010**, *15*, 13. [CrossRef]

54. Collier, N.; Campbell, B.M.; Sandker, M.; Garnett, S.T.; Sayer, J.; Boedhihartono, A.K. Science for action: The use of scoping models in conservation and development. *Environ. Sci. Policy* **2011**, *14*, 628–638. [CrossRef]

55. Frees, E.W. *Longitudinal and Panel Data: Analysis and Applications in the Social Sciences*; Cambridge University Press: Cambridge, UK, 2004.

56. Thomas, D.R. A general inductive approach for analyzing qualitative evaluation data. *Am. J. Eval.* **2006**, *27*, 237–246. [CrossRef]

57. Buys, P. *At Loggerheads?: Agricultural Expansion, Poverty Reduction, and Environment in the Tropical Forests*; World Bank Publications: Washington, DC, USA, 2007.

58. Eghenter, C.; Putera, M.H.; Ardiansyah, I. *Masyarakat Dan Konservasi 50 Kisah Yang Menginspirasi Dari WWF Untuk Indonesia*; WWF-Indonesia: Jakarta, Indonesia, 2012.

59. Wollenberg, E.; Iwan, R.; Limberg, G.; Moeliono, M.; Rhee, S.; Sudana, M. Facilitating cooperation during times of chaos: Spontaneous orders and muddling through in Malinau District, Indonesia. In *Managing Forest Resources In A Decentralized Environment*; CIFOR (Center for International Forestry Research): Bogor, Indonesia, 2007; pp. 65–74.

60. Kusters, K.; de Foresta, H.; Ekadinata, A.; Van Noordwijk, M. Towards solutions for state vs. Local community conflicts over forestland: The impact of formal recognition of user rights in Krui, Sumatra, Indonesia. *Hum. Ecol.* **2007**, *35*, 427–438. [CrossRef]

61. Thorburn, C.C. The plot thickens: Land administration and policy in post-new order Indonesia. *Asia Pac. Viewp.* **2004**, *45*, 33–49. [CrossRef]

62. Sarkar, S.; Dyer, J.S.; Margules, C.; Ciarleglio, M.; Kemp, N.; Wong, G.; Juhn, D.; Supriatna, J. Developing an objectives hierarchy for multicriteria decisions on land use options, with a case study of biodiversity conservation and forestry production from Papua, Indonesia. *Environ. Plan. B: Plan. Design* **2016**, *4*, 0265813516641684. [CrossRef]

63. Margules, C.; Sarkar, S. *Systematic Conservation Planning*; Cambridge University Press: Cambridge, UK, 2007.

64. Chomitz, K.M.; Buys, P.; Giacomo, D.L.; Timothy, S.; Sheila, W. *At Loggerheads? Agricultural Expansion, Poverty Reduction and Environment In the Tropical Forests*; World Bank: Washington, DC, USA, 2007.

65. Andersen, L.E. *The Dynamics of Deforestation and Economic Growth in the Brazilian Amazon*; Cambridge University Press: Cambridge, UK, 2002.

land

MDPI

Article

Conservation Benefits of Tropical Multifunctional Land-Uses in and Around a Forest Protected Area of Bangladesh [†]

Sharif A. Mukul [1,2,3,*] **and Narayan Saha** [4]

[1] Tropical Forestry Group, School of Agriculture and Food Sciences, The University of Queensland, Brisbane QLD 4072, Australia

[2] Tropical Forests and People Research Centre, University of the Sunshine Coast, Maroochydore QLD 4558, Australia

[3] Centre for Research on Land-use Sustainability, Noakhali 3800, Bangladesh

[4] Department of Forestry and Environmental Science, School of Agriculture and Mineral Sciences, Shahjalal University of Science and Technology, Sylhet 3114, Bangladesh; nsaha12010-fes@sust.edu

[*] Correspondence: s.mukul@uq.edu.au or sharif_a_mukul@yahoo.com; Tel.: +61-041-6648-544

[†] This paper is extended from the version presented at the 23rd IUFRO World Congress (Forests for the Future: Sustaining Society and the Environment) held in Seoul, Republic of Korea during 23–28 August 2010.

Academic Editors: Jeffrey Sayer and Chris Margules
Received: 16 November 2016; Accepted: 26 December 2016; Published: 1 January 2017

Abstract: Competing interests in land for agriculture and commodity production in tropical human-dominated landscapes make forests and biodiversity conservation particularly challenging. Establishment of protected areas in this regard is not functioning as expected due to exclusive ecological focus and poor recognition of local people's traditional forest use and dependence. In recent years, multifunctional land-use systems such as agroforestry have widely been promoted as an efficient land-use in such circumstances, although their conservation effectiveness remains poorly investigated. We undertake a rapid biodiversity survey to understand the conservation value of four contrasting forms of local land-use, namely: betel leaf (*Piper betle*) agroforestry; lemon (*Citrus limon*) agroforestry; pineapple (*Ananas comosus*) agroforestry; and, shifting cultivation–fallow managed largely by the indigenous communities in and around a highly diverse forest protected area of Bangladesh. We measure the alpha and beta diversity of plants, birds, and mammals in these multifunctional land-uses, as well as in the old-growth secondary forest in the area. Our study finds local land-use critical in conserving biodiversity in the area, with comparable biodiversity benefits as those of the old-growth secondary forest. In Bangladesh, where population pressure and rural people's dependence on forests are common, multifunctional land-uses in areas of high conservation priority could potentially be used to bridge the gap between conservation and commodity production, ensuring that the ecological integrity of such landscapes will be altered as little as possible.

Keywords: biodiversity conservation; agroforestry; traditional land-use; land-sparing; land-sharing; wildlife

1. Introduction

The vast majority of tropical forests have either been transformed or degraded by human activity, with agricultural expansion being widely recognised as the major driver of this change [1,2]. In the tropics, the ever-growing demand for land to accommodate production systems, while conserving biodiversity, providing ecosystem services, and maintaining rural people's livelihoods, make land management particularly challenging [3–6]. As forest loss and degradation continue to rise in most parts of the tropics, the international community is faced with the challenge of finding strategies that are convenient to both rural livelihoods and biodiversity conservation [7,8]. Land-sparing (setting aside land for conservation and agriculture separately) and land-sharing (integrated approach that makes land-use more conducive to biodiversity conservation) have emerged in recent years as contrasting strategies to tackle the trade-offs between livelihoods and biodiversity conservation [9]. Although the establishment of protected areas has widely been viewed as the most effective strategy to conserve biodiversity, many forest protected areas in the developing tropics are isolated and situated in a complex mosaic of agricultural land-use with high dependency of local people on them, reducing their capacity to maintain the biological diversity that they were originally designated to protect [10,11].

Agroforestry is a multifunctional land-use that involves integration of agricultural and forestry production systems in the same unit of land [8]. In the tropics, agroforestry has widely been promoted for the conservation of biodiversity with support for rural livelihoods [12]. The coffee, cacao, or the jungle rubber agroforestry systems are a few of the examples, although rapid intensification to increase crop yields and productivity make their role questionable in some regions [12,13]. Conservation biologists are also devoting an increasing amount of energy to exploring whether these multifunctional land-use systems are favourable for biodiversity conservation at both local and global scales.

Being situated in a tropical monsoon climate, Bangladesh is exceptionally rich in biodiversity [14]. Many agroforestry systems are common in the country and are managed by local and indigenous communities [15,16]. However, as with other tropical countries, intensified management of local agroforestry systems due to high demand for food and other products that has been further exacerbated by market forces and rapid agricultural development makes their role in conservation debatable [17]. Few studies, however, have so far been conducted in different parts of the country with respect to various aspects of agroforestry (e.g., [18–22]); their conservation benefit and/or ability to complement the forests have rarely been evaluated. Our study thus aimed to explore whether the multifunctional agroforestry land-use in the country is capable of biodiversity conservation (objective 1), and if it can complement the old-growth and/or less-disturbed forests (objective 2). We performed the study in a northeastern protected area of the country marked by diverse plants and wildlife. Our study is useful to understand the role of multifunctional agroforestry land-use in conservation of biodiversity in tropical human-dominated landscapes.

2. Study Area

We conducted our study in and around Lawachara National Park (LNP), one of the richest forest patches in Bangladesh (Figure 1). Several indigenous communities also live in and adjacent to the park and are dependent on it for sustaining their livelihood [16]. Geographically, the park is located between 24°30′–24°32′ N latitude and 91°37′–91°39′ E longitude with an area of about 1250 hectares. The topography of the area is undulating, with slopes and hillocks ranging from 10 to 50 m in elevation [23]. The forest of the park area originally supported tropical semi-evergreen to wet evergreen forests [24]. Presently, the area is surrounded by a complex mosaic of landscapes, dominated by tea (*Camellia sinensis*) and rubber (*Hevea brasiliensis*) gardens, plantations of commercially valuable timber species, bamboo and rattan plantations, and agricultural fields [25]. The area also experiences one of the highest rainfalls (~4000 mm/year) in the country due to its geographic location [26].

In LNP, most of the primary forest has been removed or substantially altered, with some plantations that were established in the area between 1920 and 1950 having now become part of the original forest cover and considered as old-growth secondary forest [25]. Approximately 130 ha of forest area have been used for betel leaf (*Piper betle*) cultivation within the park area, primarily by the indigenous *Khasia* community [16]. The main crop is betel leaf, a perennial dioecious climber that creeps up forest trees using its adventitious roots. This vine grows well in moist forest conditions with high humidity and soil moisture. Other forms of land-use, such as lemon agroforestry and pineapple agroforestry, are practiced mostly by the indigenous *Tripura* community and are common within and on the periphery of the park. Both lemon and pineapple agroforestry land-use require intensive management by the farmers, where they retain some of the forest tree species and cultivate few fruit species (e.g., Jackfruit; *Artocarpus heterophyllus*) with the main crop lemon in the lemon agroforestry land-use. In pineapple agroforestry land-use, farmers usually retain tree species that are tall enough or with very little branch to allow maximum sunlight to the main crop pineapple. Removal of weeds and application of chemical and organic fertilisers are also common management practices in both agroforestry land-uses.

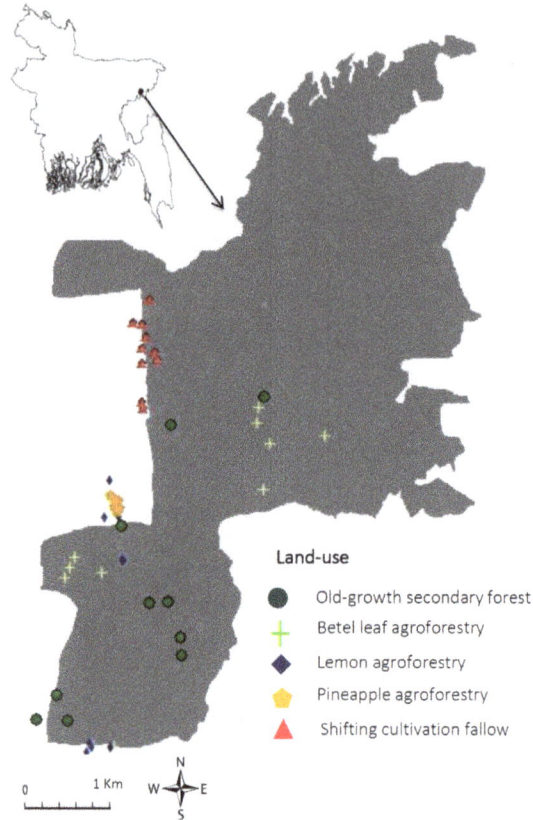

Figure 1. Map of the study area with location of the study plots used for vegetation survey in and around Lawachara National Park (LNP), Bangladesh.

While shifting cultivation is not a common form of land-use in the area, slashing and burning of an old teak plantation administered by the Forest Department and short-term agricultural use of the cleared area by the indigenous *Garo* people gave us the opportunity to study this traditional land-use common in most of the tropics [27]. We, however, acknowledge that it may not properly reflect the ideal state of shifting cultivation–fallow common in the tropics. Figure 2 shows the common land-uses in and around Lawachara National Park, Bangladesh.

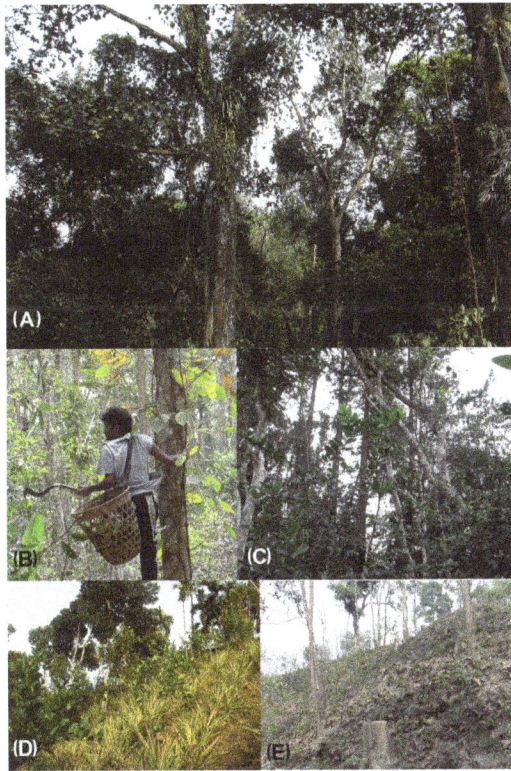

Figure 2. Different land-uses in and around Lawachara National Park—(**A**) Old-growth secondary forest; (**B**) Betel leaf agroforestry; (**C**) Lemon agroforestry; (**D**) Pineapple agroforestry; and (**E**) Shifting cultivation–fallow (Photo credits: Sharif A. Mukul).

3. Data Collection

Biodiversity assessment was undertaken during 2009 through a series of field surveys in the area. We recorded abundance and richness of trees, birds and mammals during the time of the survey. For the vegetation survey, a total of 50,100 m² randomly selected plots were established, representing four multifunctional agroforestry land-uses and the old-growth secondary forest (Figure 1). Within each 10 m × 10 m rectangular plot, we identified all mature trees ≥ 5 cm at diameter at breast height (dbh). Tree species were identified in the field and were cross-checked using available references [28,29]. For bird diversity, we used 25 points (5 land-use × 5 replicate) with a 25 m radius for a period of one hour. In each point, observations were made from a suitable place during daylight hours with periodic movements within the area to detect and identify available cryptic and non-vocal species. Nocturnal birds were excluded from this survey. For mammalian diversity, we organized a walk along 25 pre-established trails (5 land-use × 5 replicate) at a very slow pace (~2 km/h), as done by Carrillo et al. [30]. All walks were performed early in the morning and/or late in the afternoon.

4. Data Analysis

We measured alpha (α) and beta (β) diversity of trees, birds and mammals in the studied land-use both at the plot and landscape levels. Species abundance, richness and the Shannon-Wiener index (*H*) were used as measures of alpha diversity, while we used Jaccard's similarity matrix (*I*) and ordination using non-metric multi-dimensional scaling (NMDS) to determine the beta diversity of species across contrasting land-uses and plots.

Species abundance was defined as the number of individuals found in a land-use/plot. Species richness was defined as the number of unique species per land-use/plot. The Shannon-Wiener index was calculated using the equation below [31].

$$H = -\sum p_i \, ln \, p_i \qquad (1)$$

where H is the Shannon-Wiener index, p_i is the proportion of individuals found in the i-th species.

We used Bray-Curtis distance with 1000 iterations for non-metric multi-dimensional scaling using the species richness data at the plot level. Jaccard's similarity matrix was used to determine how similar or different the trees, birds or mammal species were between any pair of land-use types, and the measure was obtained using the following equation;

$$I = s_{ij}/(s_i + s_j - s_{ij}) \qquad (2)$$

where s_{ij} is the number of species found in land-use i and j, s_i is the species found in land-use i, and s_j is the species found in land-use j. The estimate ranges between 0 and 1, where the higher values indicate more similarities between two different land-uses in terms of species richness.

All statistical analysis was performed in the R Statistical environment [32]. We used the package "vegan" for diversity analysis. ANOVA (analysis of variance) and post-hoc analysis were performed to evaluate if there was any significant difference among means.

5. Results

5.1. Does Multifunctional Agroforestry Land-Use Conserve Forest Biodiversity?

During our biodiversity survey, altogether we found 44 species of trees, 53 bird species, and 19 mammal species from our sites in the Lawachara forest.
tab:land-06-00002-t001 shows the number of trees, birds and mammals found at the landscape level from each land-use type in the area. Tree species richness was highest in the betel leaf agroforestry system, while bird and mammal species richness was highest in the old-growth forest. There were no mature trees in several sites devoted to pineapple agroforestry and shifting cultivation–fallow sites in the area. These numbers, however, do not necessarily reflect the actual biodiversity of the area in studied land-use, which can be also affected by sampling design, survey area, and survey time.

Table 1. Biodiversity of trees, birds and mammals in contrasting land-uses in and around Lawachara National Park, Bangladesh.

Land-Use	Biodiversity		
	Tree *	Bird	Mammal
Old-growth secondary forest	20 (55)	31 (59)	25 (44)
Betel leaf agroforestry	22 (55)	23 (32)	16 (22)
Lemon agroforestry	8 (32)	14 (31)	15 (17)
Pineapple agroforestry	11 (10)	9 (18)	7 (9)
Shifting cultivation–fallow	9 (26)	11 (28)	16 (11)

* Values in the parenthesis indicate the abundance of species.

Figure 3 shows the biodiversity (richness and Shannon-Wiener index) of trees, birds and mammals in studied land-use types in and around Lawachara. Interestingly, both tree species abundance ($p < 0.0$; $F = 12.99$), richness ($p < 0.0$; $F = 10.45$) and the Shannon-Wiener index ($p < 0.0$; $F = 10.45$) were significantly high in the betel leaf agroforestry system, followed by the old-growth secondary forest, lemon agroforestry, shifting cultivation–fallow and pineapple agroforestry systems.

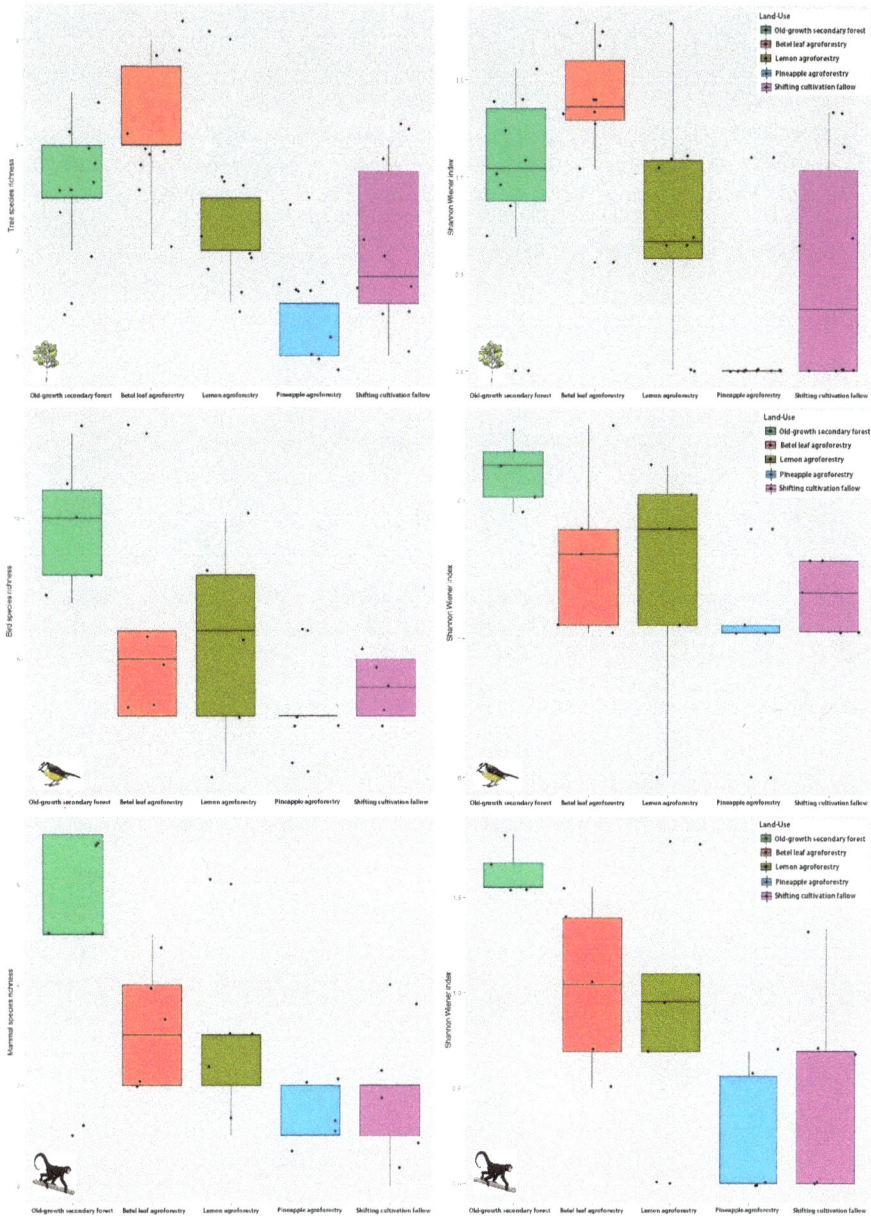

Figure 3. Biodiversity of trees, birds and mammals in contrasting land-uses in and around Lawachara National Park, Bangladesh.

Bird species abundance ($p < 0.05$; $F = 4.32$), richness ($p < 0.05$; $F = 4.05$), and the Shannon-Wiener index ($p < 0.05$; $F = 2.91$) were however significantly high in the old-growth secondary forest, followed by betel leaf agroforestry, lemon agroforestry, shifting cultivation–fallow, and pineapple agroforestry. Mammal abundance ($p < 0.01$; $F = 4.97$) and richness ($p < 0.05$; $F = 3.63$) were significantly high in the old-growth secondary forests, while there were no significant differences in the Shannon-Wiener index ($p > 0.05$; $F = 2.73$) across the land-uses.

5.2. Can Tropical Multifunctional Land-Uses Complement the Less-Disturbed Forest?

Tables 2–4 show the similarities of tree, bird and mammal species richness across the landscapes in different agroforestry land-uses as well as in the old-growth secondary forest in the Lawachara area using Jaccard's matrix of similarities. Interestingly, the old-growth secondary forest shared higher numbers of species with lemon agroforestry and shifting cultivation–fallow in the area than betel leaf agroforestry land-use. Betel leaf agroforestry land-use, however, had the highest number of common birds and mammals with the old-growth secondary forest than other land-uses in the area (see Tables 3 and 4).

Table 2. Similarity matrix of tree species across the studied land-uses in and around Lawachara.

Land-Use	Betel Leaf Agroforestry	Lemon Agroforestry	Pineapple Agroforestry	Shifting Cultivation–Fallow
Old-growth secondary forest	0.11	0.30	0.09	0.17
Betel leaf agroforestry		0.24	0.13	0.11
Lemon agroforestry			0.26	0.13
Pineapple agroforestry				0

Table 3. Similarity matrix of bird species across the studied land-uses in and around Lawachara.

Land-Use	Betel Leaf Agroforestry	Lemon Agroforestry	Pineapple Agroforestry	Shifting Cultivation–Fallow
Old-growth secondary forest	0.29	0.16	0.08	0.14
Betel leaf agroforestry		0.19	0.07	0.22
Lemon agroforestry			0.21	0.14
Pineapple agroforestry				0.11

Table 4. Similarity matrix of mammals across the studied land-uses in and around Lawachara.

Land-Use	Betel Leaf Agroforestry	Lemon Agroforestry	Pineapple Agroforestry	Shifting Cultivation–Fallow
Old-growth secondary forest	0.53	0.53	0.47	0.26
Betel leaf agroforestry		0.47	0.64	0.46
Lemon agroforestry			0.50	0.46
Pineapple agroforestry				0.36

When looking at plot level similarities between different land-use in the area using non-metric multi-dimensional scaling and species richness, we observed greater overlaps of tree species between the old-growth secondary forest and lemon agroforestry. NMDS reveals overlaps of tree species with different land-uses and the old-growth secondary forest in the area (Stress = 14.0). A heterogeneous bird species assemblage was found across the plots of different land-use (Stress = 20.58), although a greater number of bird species was shared between betel leaf agroforestry and the old-growth secondary forest in the area. All agroforestry land-use types shared some mammal species with the old-growth secondary forest in the area, although we observed no visible pattern in mammal species assemblage in the area (Stress = 17.75). Figures 4–6 show ordination of plots using species richness of mature trees, birds and mammals in the area, where scattered plots indicate a more heterogeneous species assemblage, and closer plots indicate a more homogenous species assemblage across plots of different land-use types. The axes here in the figures are arbitrary as is the orientation of the plot.

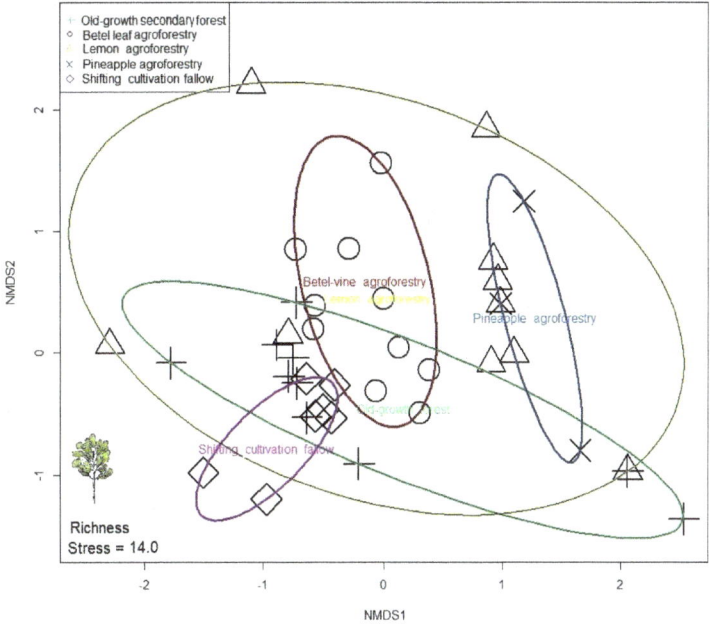

Figure 4. Ordination using non-metric multi-dimensional scaling and tree species richness.

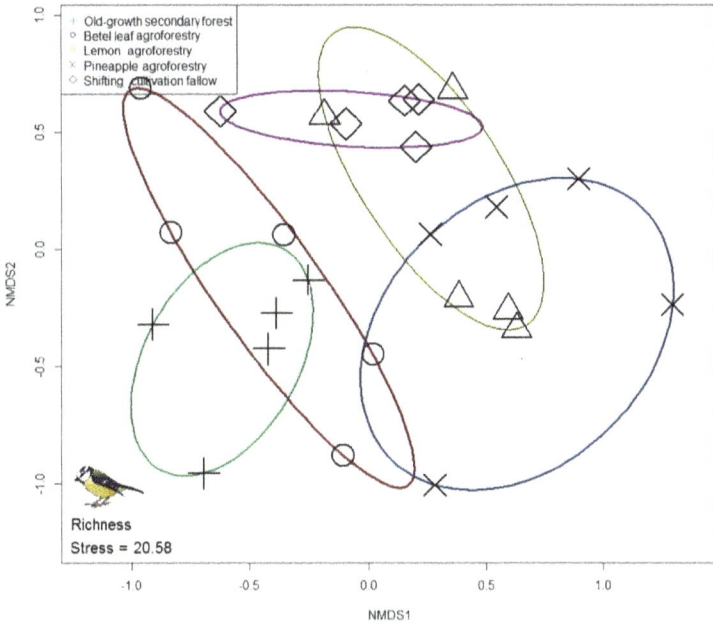

Figure 5. Ordination using non-metric multi-dimensional scaling and bird species richness.

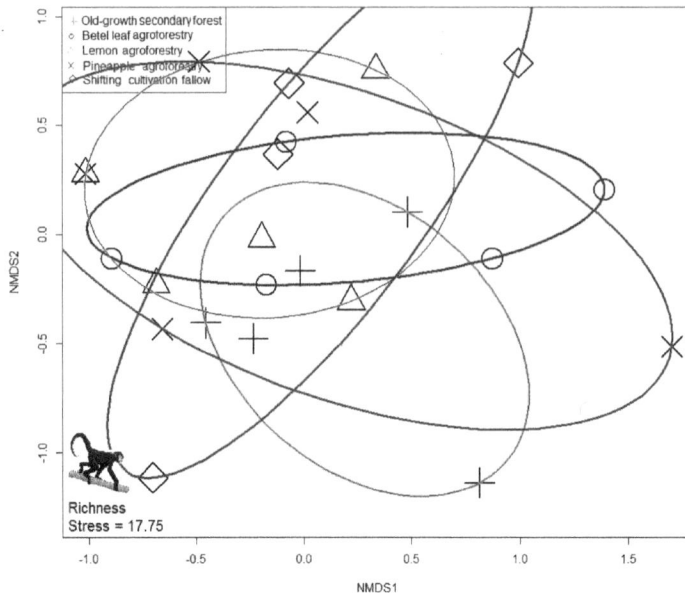

Figure 6. Ordination using non-metric multi-dimensional scaling and mammal species richness.

6. Discussion and Conclusions

We find tropical multifunctional agroforestry land-use suitable for biodiversity conservation in and around the Lawachara forest, although the ability of different land-use types to protect and conserve biodiversity was not similar. In our study, we observed a higher diversity of tree in betel leaf agroforestry, although bird and mammal diversity were higher in the old-growth secondary forest in the area. Compared to other land-use types, pineapple agroforestry, and shifting cultivation–fallow had a lower diversity of plants and mammals in the area. The betel leaf agroforestry land-use managed by the indigenous *Khasia* community provides superior conservation benefits in the area, which may be because it involved relatively less modification of natural vegetation compared to other land-uses [15]. Smallholder landowners here retain the native forest trees to use them as support for the betel leaf vine. The pineapple agroforestry system, on the other hand, involves the clearing of native vegetation to allow greater sunlight in land under use, with regular human intervention and the use of agrochemicals. Lemon agroforestry, although not as intensive as the pineapple agroforestry system, also involved more active management of the land for lemon production. When considering the ability of different multifunctional land-use types to complement the native vegetation of old-growth and/or less-disturbed secondary forests, we find betel leaf agroforestry also suitable for forest bird and mammal species, although lemon agroforestry shared a higher number of tree species with the old-growth secondary forest in the area than other agroforestry land-uses. We, however, did not include some other factors such as proximity to less-disturbed forest, the age of the land-use in our analysis, which may also influence the biodiversity in the studied land-use types [26]. For instance, the limited biodiversity in our shifting cultivation–fallow sites can be due to the relatively young age of the sites and/or recent history of disturbance. Tropical primary forests are believed to be irreplaceable for their unique biodiversity [33]. In human-dominated tropical forest landscapes, the increasing pressure on land for agriculture, however, makes their management complex [34]. In such a context, wildlife-friendly agriculture has been widely advocated to reduce the adverse effect of agriculture [35,36]. In many parts of the tropics, agroforestry land-uses have been common for centuries and making a contribution as a refuge for biodiversity conservation [8,13,37]. In recent years, rapid intensification of land-use with excessive use of agrochemicals and modification of land for higher yields has been unfavorably affecting the conservation value of such landscapes [17].

In Bangladesh, biodiversity conservation is challenging due to high population pressure, limited land designated for conservation, and high dependence of rural people on forests [25,38]. Conflicts between protected area managers and local peoples are common in the country due to limited access of local people to different income-generating options and poor recognition of their traditional ways of living [16,25]. In tropical multifunctional landscapes, there is a strong linkage between income and environmental benefits that sometime influence smallholders' motivation and preference for particular land-use(s) [39–41]. Earlier studies also suggest that, to some extent, higher protection status and local beliefs help to protect the declining biodiversity of the country [26,42,43]. The present study demonstrates that multifunctional land-uses such as agroforestry could potentially be used to protect and conserve biodiversity in the country, ensuring limited modification of the native vegetation and wildlife habitats. Such land-uses also hold great promise for integration in the available carbon forestry schemes, including reducing emissions from deforestation and forest degradation (REDD+) with benefits to both local livelihoods and the environment, although further studies are essential to quantify such benefits [44].

Acknowledgments: We would like to thank our respondents for sharing the information with us and Mashiur Rahman Tito, Mofizul Haque, Mohammad Abu Sayed Arfin Khan for accompanying during the study. The authors would like to acknowledge the infrastructure and other support provided by the Centre for Research on Land-use Sustainability, Bangladesh. The first author (SAM) was supported by the British Ecological Society (SEPG, UEPS Field Experience Grant 7210 600), British Ornithologist's Union (2009) and the Rufford Grant for Nature Conservation (RSG project 69.01.09). The authors greatly acknowledge the useful comments and suggestions from three anonymous reviewers.

Author Contributions: S.A.M. conceived the idea, conducted the field work and analysed the data; S.A.M. and N.S. wrote the paper.

Conflicts of Interest: The authors declare no conflicts of interest.

References

1. Uriarte, M.; Schneider, L.; Rudel, T.K. Land transitions in the tropics: Going beyond the case studies. *Biotropica* **2010**, *42*, 1–2. [CrossRef]

2. Sala, O.E.; Chapin, F.S.; Amesto, J.J. Global biodiversity scenarios for the year 2100. *Science* **2000**, *287*, 1770–1774. [CrossRef] [PubMed]

3. Law, E.A.; Bryan, B.A.; Meijaard, E.; Mallawaarachchi, T.; Struebig, M.J.; Watts, M.E.; Wilson, K.A. Mixed policies give more options in multifunctional tropical forest landscapes. *J. Appl. Ecol.* **2016**. [CrossRef]

4. Mukul, S.A.; Herbohn, J.; Firn, J. Co-benefits of biodiversity and carbon sequestration from secondary forests in the Philippine uplands: Implications for forest landscape restoration. *Biotropica* **2016**, *48*, 882–889. [CrossRef]

5. Sayer, J.; Sunderland, T.; Ghazoul, J.; Pfund, J.L.; Sheil, D.; Meijaard, E.; Venter, M.; Boedhihartono, A.K.; Day, M.; Garcia, C. Ten principles for a landscape approach to reconciling agriculture, conservation, and other competing land uses. *Proc. Natl. Acad. Sci. USA* **2013**, *110*, 8349–8356. [CrossRef] [PubMed]

6. Harvey, C.A.; Gonzalez, J.A. Agroforestry systems conserve species-rich but modified assemblages of tropical birds and bats. *Biodivers. Conserv.* **2007**, *16*, 2257–2292. [CrossRef]

7. Santika, T.; Meijaard, E.; Wilson, K.A. Designing multifunctional landscapes for forest conservation. *Environ. Res. Lett.* **2015**, *10*, 114012. [CrossRef]

8. Bhagwat, S.A.; Willis, K.J.; Birks, H.J.B.; Whittaker, R.J. Agroforestry: A refuge for tropical biodiversity? *Trends Ecol. Evolut.* **2008**, *23*, 261–267. [CrossRef] [PubMed]

9. Kremen, C. Reframing the land-sparing/land-sharing debate for biodiversity conservation. *Ann. N. Y. Acad. Sci.* **2015**, *1355*, 52–76. [CrossRef] [PubMed]

10. Mukul, S.A.; Rashid, A.Z.M.M.; Uddin, M.B.; Khan, N.A. Role of non-timber forest products in sustaining forest-based livelihoods and rural households' resilience capacity in and around protected area: A Bangladesh study. *J. Environ. Plan. Manag.* **2016**, *59*, 628–642. [CrossRef]

11. Nagendra, H. Do parks work? Impact of protected areas on land cover clearing. *Ambio* **2008**, *37*, 330–337. [CrossRef] [PubMed]

12. Schroth, G.; da Fonseca, G.A.B.; Harvey, C.A.; Gascon, C.; Vasconcelos, H.L.; Izac, A.N. (Eds.) *Agroforestry and Biodiversity Conservation in Tropical Landscapes*; Island Press: Washington, DC, USA, 2004.

13. Steffan-Dewenter, I.; Kessler, M.; Barkmann, J.; Bos, M.M.; Buchori, D.; Erasmi, S.; Faust, H.; Gerold, G.; Glenk, K.; Gradstein, S.R.; et al. Tradeoffs between income, biodiversity, and ecosystem functioning during tropical rainforest conversion and agroforestry intensification. *Proc. Natl. Acad. Sci. USA* **2007**, *104*, 4973–4978. [CrossRef] [PubMed]

14. Mukul, S.A.; Rashid, A.Z.M.M.; Khan, N.A. Forest protected area systems and biodiversity conservation in Bangladesh. In *Protected Areas: Policies, Management and Future Directions*; Mukul, S.A., Rashid, A.Z.M.M., Eds.; Nova Science Publishers: New York, NY, USA, 2017; pp. 157–177.

15. Mukul, S.A. Biodiversity conservation and ecosystem functions of traditional agroforestry systems: Case study from three tribal communities in and around Lawachara National Park. In *Forest Conservation in Protected Areas of Bangladesh: Policy and Community Development Perspectives*; Chowdhury, M.S.H., Ed.; Springer: Basel, Switzerland, 2014; pp. 171–179.

16. Mukul, S.A. The role of traditional forest practices in enhanced conservation and improved livelihoods of indigenous communities: Case study from Lawachara National Park, Bangladesh. In Proceedings of the 1st International Conference on Forest Related Traditional Knowledge and Culture in Asia, Seoul, Korea, 5–10 October 2008; pp. 24–28.

17. Mukul, S.A. Ecological Trade-Offs between Agroforestry Land-Use, Biodiversity Conservation and Management Intensification, Case Study from in and Around Lawachara National Park, Moulvibazar, Bangladesh. Master's Thesis, Shahjalal University of Science and Technology, Sylhet, Bangladesh, 2009.

18. Mukul, S.A.; Tito, M.R.; Munim, S.A. Can homegardens help save forests in Bangladesh? Domestic biomass fuel consumption patterns and implications for forest conservation in south-central Bangladesh. *Int. J. Res. Land-Use Sustain.* **2014**, *1*, 18–25.

19. Kabir, M.E.; Webb, E.L. Can homegardens conserve biodiversity in Bangladesh? *Biotropica* **2008**, *40*, 95–103. [CrossRef]

20. Alam, M.K.; Ahmed, F.U.; Ruhul-Amin, S.M. (Eds.) *Agroforestry: Bangladesh Perspective*; Bangladesh Agricultural Research Council: Dhaka, Bangladesh, 2007.

21. Saha, N.; Azam, M.A. The indigenous hill-farming system of *Khasia* tribes in Moulvibazar district of Bangladesh: Status and impacts. *Small-Scale For. Econ. Manag. Policy* **2004**, *3*, 273–281.

22. Khan, N.A.; Khisa, S.K. Sustainable land management with rubber-based agroforestry: A Bangladeshi example of uplands community development. *Sustain. Dev.* **2000**, *8*, 1–10. [CrossRef]

23. Sohel, M.S.I.; Mukul, S.A.; Burkhard, B. Landscape's capacities to supply ecosystem services in Bangladesh: A mapping assessment for Lawachara National Park. *Ecosyst. Serv.* **2015**, *12*, 128–135. [CrossRef]

24. MacKinnon, J.R. *Protected Areas Systems: Review of the Indo-Malayan Realm*; The Asian Bureau for Conservation Limited: Canterbury, UK, 1997.

25. Mukul, S.A.; Herbohn, J.; Rashid, A.Z.M.M.; Uddin, M.B. Comparing the effectiveness of forest law enforcement and economic incentive to prevent illegal logging in Bangladesh. *Int. For. Rev.* **2014**, *16*, 363–375. [CrossRef]

26. Pavel, M.A.A.; Mukul, S.A.; Uddin, M.B.; Harada, K.; Khan, M.A.S.A. Effects of stand characteristics on tree species richness in and around a conservation area of northeast Bangladesh. *J. Mt. Sci.* **2016**, *13*, 1085–1095. [CrossRef]

27. Mukul, S.A.; Herbohn, J. The impacts of shifting cultivation on secondary forests dynamics in tropics: A synthesis of the key findings and spatio temporal distribution of research. *Environ. Sci. Policy* **2016**, *55*, 167–177. [CrossRef]

28. Dey, T.K. *Useful Plants of Bangladesh*; Add Communication: Chittagong, Bangladesh, 2006. (In Bengali)

29. Das, D.K.; Alam, M.K. Trees of Bangladesh. In *Bangladesh Forest Research Institute*; Bangladesh Forest Research Institute (BFRI): Chittagong, Bangladesh, 2001.

30. Carrillo, E.; Wong, G.; Cuarón, A.D. Monitoring mammal populations in Costa Rican protected areas under different hunting restrictions. *Conserv. Biol.* **2000**, *14*, 1580–1591. [CrossRef]

31. Magurran, A.E. *Measuring Biological Diversity*; Blackwell Publishing Company: Oxford, UK, 2004.

32. R Development Core Team. *R: A Language and Environment for Statistical Computing*; R Foundation for Statistical Computing: Vienna, Austria, 2015.

33. Gibson, L.; Lee, T.M.; Koh, L.P.; Brook, B.W.; Gardner, T.A.; Barlow, J.; Peres, C.A.; Bradshaw, C.J.A.; Laurance, W.F.; Lovejoy, T.E.; et al. Primary forests are irreplaceable for sustaining tropical biodiversity. *Nature* **2011**, *478*, 378–381. [CrossRef] [PubMed]

34. DeFries, R.S.; Foley, J.A.; Asner, G.P. Land-use choices: Balancing human needs and ecosystem function. *Front. Ecol. Environ.* **2004**, *2*, 249–257. [CrossRef]

35. Harvey, C.A.; Komar, O.; Chazdon, R.; Ferguson, B.G.; Finegan, B.; Griffith, D.M.; Martinez-Ramos, M.; Morales, H.; Nigh, R.; Soto-Pinto, L.; et al. Integrating agricultural landscapes with biodiversity conservation in the Mesoamerican hotspot. *Conserv. Biol.* **2008**, *22*, 8–15. [CrossRef] [PubMed]

36. Green, R.E.; Cornell, S.J.; Scharlemann, J.P.W.; Balmford, A. Farming and the fate of wild nature. *Science* **2005**, *307*, 550–555. [CrossRef] [PubMed]

37. Moguel, P.; Toledo, V.M. Biodiversity conservation in traditional coffee systems of Mexico. *Conserv. Biol.* **1999**, *13*, 11–21. [CrossRef]

38. Mukul, S.A.; Uddin, M.B.; Rashid, A.Z.M.M.; Fox, J. Integrating livelihoods and conservation in protected areas: Understanding role and stakeholders' views on the prospects of non-timber forest products, A Bangladesh case study. *Int. J. Sustain. Dev. World Ecol.* **2010**, *17*, 180–188. [CrossRef]

39. Saha, N.; Azam, M.A. Betel leaf based forest farming by khasia tribes: A sustainable system of forest management in Moulvibazar district, Bangladesh. *For. Trees Livelihoods* **2005**, *5*, 275–290. [CrossRef]

40. Rahman, S.A.; Sunderland, T.; Kshatriya, M.; Roshetko, J.M.; Pagella, T.; Healey, J.R. Towards productive landscapes: Trade-offs in tree-cover and income across a matrix of smallholder agricultural land-use systems. *Land Use Policy* **2016**, *58*, 152–164. [CrossRef]

41. Clough, Y.; Krishna, V.V.; Corre, M.D.; Darras, K.; Denmead, L.H.; Meijide, A.; Moser, S.; Musshoff, O.; Steinebach, S.; Veldkamp, E.; et al. Land-use choices follow profitability at the expense of ecological functions in Indonesian smallholder landscapes. *Nat. Commun.* **2016**, *7*, 13137. [CrossRef] [PubMed]

42. Mukul, S.A.; Rashid, A.Z.M.M.; Uddin, M.B. The role of spiritual beliefs in conserving wildlife species in religious shrines of Bangladesh. *Biodiversity* **2012**, *13*, 108–114. [CrossRef]

43. Uddin, M.B.; Steinbauer, M.J.; Jentsch, A.; Mukul, S.A.; Beierkuhnlein, C. Do environmental attributes, disturbances, and protection regimes determine the distribution of exotic plant species in Bangladesh forest ecosystem? *For. Ecol. Manag.* **2013**, *303*, 72–80. [CrossRef]

44. Mukul, S.A.; Biswas, S.R.; Rashid, A.Z.M.M.; Miah, M.D.; Kabir, M.E.; Uddin, M.B.; Alamgir, M.; Khan, N.A.; Sohel, M.S.I.; Chowdhury, M.S.H.; et al. A new estimate of carbon for Bangladesh forest ecosystems with their spatial distribution and REDD+ implications. *Int. J. Res. Land-Use Sustain.* **2014**, *1*, 33–41.

Article

Large-Scale Mapping of Tree-Community Composition as a Surrogate of Forest Degradation in Bornean Tropical Rain Forests

Shogoro Fujiki [1,*], **Ryota Aoyagi** [1], **Atsushi Tanaka** [1,2], **Nobuo Imai** [1,3], **Arif Data Kusma** [4], **Yuyun Kurniawan** [5], **Ying Fah Lee** [6], **John Baptist Sugau** [6], **Joan T. Pereira** [6], **Hiromitsu Samejima** [7] **and Kanehiro Kitayama** [1]

[1] Graduate School of Agriculture, Kyoto University, Kyoto 606-8502, Japan; aoyagi.ryota@gmail.com (R.A.); amcw2003@gmail.com (A.T.); i96nobuo@gmail.com (N.I.); kanehiro@kais.kyoto-u.ac.jp (K.K.)
[2] Japan Forest Technology Association, Tokyo 102-0085, Japan
[3] Primate Research Institute, Kyoto University, Aichi 484-8506, Japan
[4] WWF Indonesia, Kutai Barat, Kalimantan Timur, Indonesia; AKusuma@wwf.or.id
[5] WWF Indonesia, Ujung Kulon, Banten, Indonesia; YKurniawan@wwf.or.id
[6] Forest Research Centre, Sabah Forestry Department, 90715 Sandakan, Sabah, Malaysia; yingfah.lee@sabah.gov.my (Y.F.L.); john.sugau@sabah.gov.my (J.B.S.); Joan.pereira@sabah.gov.my (J.T.P.)
[7] Institute for Global Environmental Strategies, Kanagawa 240-0115, Japan; lahang.lejau@gmail.com
* Correspondence: fujiki5636@gmail.com; Tel.: +81-75-753-6080

Academic Editors: Jeffrey Sayer and Chris Margules
Received: 27 July 2016; Accepted: 6 December 2016; Published: 11 December 2016

Abstract: Assessment of the progress of the Aichi Biodiversity Targets set by the Convention on Biological Diversity (CBD) and the safeguarding of ecosystems from the perverse negative impacts caused by Reducing Emissions from Deforestation and Forest Degradation Plus (REDD+) requires the development of spatiotemporally robust and sensitive indicators of biodiversity and ecosystem health. Recently, it has been proposed that tree-community composition based on count-plot surveys could serve as a robust, sensitive, and cost-effective indicator for forest intactness in Bornean logged-over rain forests. In this study, we developed an algorithm to map tree-community composition across the entire landscape based on Landsat imagery. We targeted six forest management units (FMUs), each of which ranged from 50,000 to 100,000 ha in area, covering a broad geographic range spanning the most area of Borneo. Approximately fifty 20 m-radius circular plots were established in each FMU, and the differences in tree-community composition at a genus level among plots were examined for trees with diameter at breast height $\geq$10 cm using an ordination with non-metric multidimensional scaling (nMDS). Subsequently, we developed a linear regression model based on Landsat metrics (e.g., reflectance value, vegetation indices and textures) to explain the nMDS axis-1 scores of the plots, and extrapolated the model to the landscape to establish a tree-community composition map in each FMU. The adjusted R^2 values based on a cross-validation approach between the predicted and observed nMDS axis-1 scores indicated a close correlation, ranging from 0.54 to 0.69. Histograms of the frequency distributions of extrapolated nMDS axis-1 scores were derived from each map and used to quantitatively diagnose the forest intactness of the FMUs. Our study indicated that tree-community composition, which was reported as a robust indicator of forest intactness, could be mapped at a landscape level to quantitatively assess the spatial patterns of intactness in Bornean rain forests. Our approach can be used for large-scale assessments of tree diversity and forest intactness to monitor both the progress of Aichi Biodiversity Targets and the effectiveness of REDD+ biodiversity safeguards in production forests in the tropics.

Keywords: Aichi Biodiversity Targets; environmental safeguards; forest intactness; logged-over forests; REDD+; satellite remote sensing; tropical production forests

1. Introduction

Continued deforestation and forest degradation and the associated losses of biodiversity in tropical countries represent major global concerns [1]. To date, coordinated international efforts have resulted in two international conventions that attempt to reduce the rate of tropical deforestation and forest degradation and the associated biodiversity losses: the Convention on Biological Diversity (CBD), and the United Nations Framework Convention on Climate Change (UNFCCC).

The CBD sets a strategic plan for biodiversity for 2011–2020 and the Aichi Biodiversity Targets, which include several targets for forest conservation and their sustainable use [2]. It is important to quantitatively assess the progress of the Aichi Biodiversity Targets, and there has been intensive discussion on which indicators should be used for monitoring them [3,4]. The development of essential biodiversity variables as a measurement for studying, reporting, and managing changes in biodiversity is a prerequisite for achieving the targets [4].

The UNFCCC is a convention primarily targeting the mitigation of and adaptation to climate change. Reducing Emissions from Deforestation and Forest Degradation Plus (REDD+), a post-Kyoto Protocol mechanism developed under UNFCCC, is expected to have positive effects on biodiversity conservation because its main targets are natural tropical forests. When considering the impact of REDD+, it is important to consider REDD+ activities with biodiversity trade-offs. For example, if low-carbon, high-biodiversity tropical forests are converted into high-carbon, low-biodiversity forests (e.g., plantations and forests dominated by pioneer species), REDD+ may have an overall negative effect on biodiversity conservation [5,6]. Measures to safeguard biodiversity have been extensively discussed in this context under UNFCCC [7–9]. The premise of these safeguards is a compliance system where each REDD+ project is required to comply with standards and indicators to qualify for REDD+ credits. Such a compliance system must involve third party auditing and verification based on standards and indicators. Therefore, there is an urgent need to develop robust indicators to both accurately assess the progress of the Aichi Biodiversity Targets and avoid overall negative effects on biodiversity caused by the implementation of REDD+. A key obstacle to achieving these goals is the lack of global, integrated observation systems for delivering regular, timely data on changes in biodiversity [4].

Indicators for compliance systems can be generic or specific. Generic indicators [10] are a set of indicators applicable to all forest types and regions; they may be largely based on readily available statistical data or actual prescribed management plans such as the presence/area of forests with high conservation value, presence/type of conservation measures, and presence/magnitude of mitigation measures. Specific indicators are a set of indicators that involve direct measurements in the field such as changes in the distribution/number of endangered species, changes in the area of intact ecosystems, and changes in species richness. For indicators to be reliable and practical, they must incur a low financial cost, be easily identified, be proxies for ecosystem integrity, and have cross-taxon congruency [11] as well as sufficient robustness and sensitivity [12]. Generic indicators are superior in cost effectiveness, while specific indicators are superior as proxies of ecosystem integrity and sensitivity.

Recently, Imai et al. [12] proposed that the community composition of canopy trees could be used as a specific indicator for "ecosystem integrity or intactness" in spatiotemporally dynamic Bornean production forests where timber is produced by commercial logging. The authors used the axis-1 scores of the ordination of vegetation plots based on the relative species (or genus) abundances in logged-over forests to indicate forest intactness. Derived axis-1 scores showed significant linear correlations with magnitude of logging intensity (i.e., the inverse of remaining above-ground biomass). This significant linear correlation was a result of the interaction of linearly increasing species number in the pioneer guild and linearly decreasing species number in the climax guild with increased logging intensity [12,13]. Therefore, the indicator based on community composition (i.e., axis-1 scores of the ordination) actually indicated the compositional distance from an intact forest where there were zero or minimal effects from logging. The count-plot measurements on the ground were fast and inexpensive, and implemented by local foresters [12]. Therefore, the methods of Imai et al. [12] satisfied the above-listed requirements for indicators.

A previously unresolved issue related to specific indicators is spatial representativeness; specific indicators derived from plots on the ground need to be extrapolated to the landscape or even region level. To address the increasing need for a practical indicator that is spatially/temporally sensitive and robust, we developed a new algorithm using satellite remote sensing to map forest intactness in Bornean tropical rain forests based on the community composition indicator of Imai et al. [12]. We used generic abundances instead of specific abundances because both specific and generic abundances demonstrated the same response to logging intensity [12]. Specifically, we developed a new algorithm to map variation in tree-community composition using Landsat imagery as a proxy for forest intactness and assessed whether our method could reliably diagnose the magnitude of forest intactness among/within production forests under different management regimes. We developed a linear regression model based on Landsat metrics to explain the community composition indicator of Imai et al. [12] and extrapolated the model to the landscape to establish a tree-community composition map in each forest management unit (FMU). Such a method will provide a valuable contribution to assessments of the progress of meeting the Aichi Biodiversity Targets and effectiveness of REDD+ safeguards as well as improved management of forest management units (FMUs).

2. Materials and Methods

2.1. Study Site

We studied the community composition of canopy trees and mapped the patterns of tree-community composition in six Bornean FMUs that were conducting legal selective logging for commercial purposes. An FMU is a management entity with a valid logging license from the local government either from Malaysian states in Malaysian Borneo or from the Indonesian Government in Indonesian Borneo. In each FMU, the marketable trees (large dipterocarp species) within a certain size class are harvested for timber. FMUs in Borneo typically range from 50,000 to 100,000 ha in size, and their land belongs to the respective governments. These areas are covered by lowland mixed dipterocarp forests, with varying degrees of degradation reflecting their different logging histories.

The studied FMUs were Segaliud Lokan (5°20′–27′N, 117°23′–39′E, 576 km^2), Deramakot (5°14′–28′N, 117°20′–38′E, 551 km^2), Tangkulap (5°18′–31′N, 117°11′–22′E, 276 km^2), and Sapulut (4°40′–55′N, 116°30′–117°00′E, 956 km^2) in Sabah, Malaysia; and, Roda Mas (0°46′–1°05′N, 114°25′–115°06′E, 703 km^2) and Ratah (0°7′S–0°13′N, 114°58′–115°30′E, 982 km^2) in East Kalimantan, Indonesia (Figure 1). The areas of Segaliud Lokan, Deramakot, and Tangkulap were initially logged in 1958, 1956, and 1970, respectively, using conventional logging methods (i.e., high-impact logging with no environmental considerations [14]). The three FMUs were adjacent to one another. In Deramakot, conventional logging continued until 1989, when all logging activities were halted for regrowth. Then, a long-term management plan with reduced-impact logging was introduced to Deramakot in 1995. Reduced-impact logging is an improved method of selective logging, including pre-harvest inventory, mapping of all canopy trees, directional felling, liana cutting, and planning of skid trails, log decks, and roads [15,16]. In combination with reduced-impact logging, a longer cutting cycle (i.e., 40 years) was strictly adhered to in accordance with the long-term management plan [17]. These combined approaches helped to preserve forest integrity [13,18–20]. Deramakot was the first tropical forest certified by the Forest Stewardship Council (FSC) in 1997, and was considered an exemplary model of sustainable forest management by the Sabah Government. Reflecting this logging history, we observed that the forests inside the Deramakot FMU were less disturbed based on aboveground biomass and varied biological communities [13]. In contrast, Segaliud Lokan was repeatedly logged using conventional logging until 2002, after which reduced-impact logging has been implemented. Tangkulap was repeatedly logged using conventional logging until 2002, after which all logging activities have remained suspended. Sapulut was first logged in 1956 and repeatedly logged until 2000 using conventional logging, after which reduced-impact logging has been carried out, but industrial tree plantations have been established and Sapulut now consists of rubber and Acacia plantations over approximately 35% of its area. Reduced-impact logging was implemented in Roda Mas from at least 2008. It is highly likely that logging had been conducted earlier in the Roda Mas area, but there is no information about its logging history before 2008. We estimate that the logging intensity was relatively mild in the Roda Mas area because areas of intact forest still remain throughout the area. Ratah implemented conventional logging from 1972 to 2010, after which reduced-impact logging was implemented. Deramakot, Tangkulap, Roda Mas, and Ratah are all certified by the FSC. Sapulut and Segaliud Lokan are certified by Malaysian Timber Certification Council (MTCC). Table 1 summarizes the target areas.

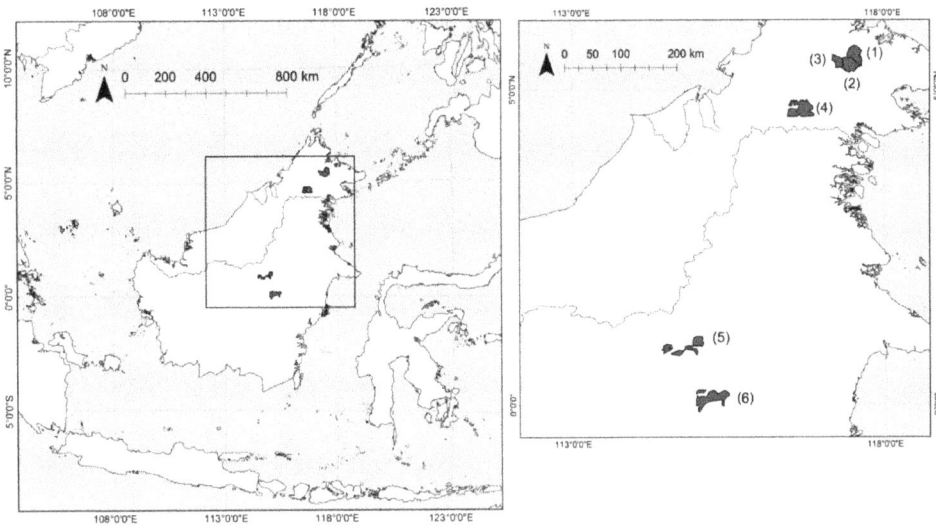

Figure 1. Locations of the six forest management units (FMUs) in this study: (1) Segaliud Lokan, (2) Deramakot, (3) Tangkulap, (4) Sapulut, (5) Roda Mas, and (6) Ratah.

Table 1. Profiles of the forest management units (FMUs) targeted in this study.

Name of FMU	State/Province, Nation	Silviculture and Timber Harvest	Forest Certification	Established Plot Number
Segaliud Lokan		1958–2002 CL, Since 2003 RIL	MTCC	
Deramakot	Sabah, Malaysia	1956–1985 CL, Since 1995 RIL	FSC	50
Tangkulap		1970–2002 CL	FSC	
Sapulut		1956–2000 CL, Since 2001 RIL	MTCC	50
Roda Mas	East Kalimantan, Indonesia	Since at least 2008 RIL	FSC	50
Ratah		1972–2010 CL, Since 2011 RIL	FSC	90

Note: CL, conventional logging; RIL, reduced impact logging; FSC, Forest Stewardship Council; MTCC, Malaysian Timber Certification Council.

2.2. Field Survey

We conducted count-plot surveys in the four FMUs: Segaliud Lokan, Sapulut, and Ratah in 2011–2012 [12] and Roda Mas and Ratah in 2014–2015. The number of the established plots in each FMU is shown in Table 1. Because Segaliud Lokan, Deramakot, and Tangkulap are adjacent to one another with the same forest type, we assumed that the count-plot data obtained from the field survey in Segaliud Lokan was representative of the canopy species composition of the forests in all three FMUs.

Details of the procedure of the count-plot survey are described in Imai et al. [12]. Briefly, to select representative vegetation plots from a heterogeneous forest with varying magnitudes of forest degradation, we classified each FMU into five strata based on Landsat imagery according to the extent of forest degradation, from intact forest (stratum 1) to open canopy area with several pioneer trees (stratum 5) [12]. Here, we defined the magnitude of forest degradation based on the remaining aboveground biomass, a definition of the Intergovernmental Panel on Climate Change [21]. Ten 20 m radius circular plots (1257 m^2 in area) were randomly established at altitudes below 600 m a.s.l. in each stratum in each FMU (i.e., 50 plots in each FMU). To minimize spatial autocorrelation, plots were established at a distance of at least 100 m from one another. Global positioning system (GPS) data were collected during the field surveys at the center of a plot with calibration for at least one hour in each plot. We measured diameter at breast height (dbh) of all trees (dbh $\geq$10 cm) and counted all measured species in each plot. Woody vines were excluded from the inventory. Trees with buttresses were measured well above (ca. 50 cm) protrusions. All trees were identified by local botanical experts. If trees could not be identified in the field, voucher specimens were collected and identified in local herbaria. Samples that could not be identified to the species level were distinguished as morphospecies. We obtained inventory data from a total of 240 plots.

2.3. Field Data Analysis

We recorded 14,783 stems from 85 families and 254 genera in 240 plots (totaling 30.2 ha in area). The Chao distances [22] and the number of trees of each genus were used to calculate the distance matrix for the inventoried plots in each FMU. Then, an ordination of plots was conducted to ordinate the inventoried plots (approximately 50 plots) for each FMU with non-metric multidimensional scaling (nMDS) using the metaMDS procedure in the vegan package in the R software program (R version 3.2.0) [23]. Subsequently, the nMDS axis-1 scores were used as an index of tree-community composition. Imai et al. [12] demonstrated that the nMDS axis-1 scores of plots correlated with the aboveground biomass values of the plots, a proxy for forest degradation, in each FMU (R^2=0.52–0.71). Although plots were a considerable distance from one another, we were unable to completely rule out the possibility of spatial autocorrelations among plots. Therefore, we tested for autocorrelations among plots in each FMU but found none (see Supplementary Materials, Figures S1 and S2). Hence, the nMDS axis-1 scores of plots indicated intactness in terms of tree-community composition—in other words, the compositional distance from intact forests where there were no effects of logging. Intactness can also be functionally translated to ecosystem integrity because regeneration ability is assured in more intact forests.

To compare the differences between FMUs with the same index, the nMDS axis-1 scores were normalized by transforming the scores into new scores with a mean of 0 and a standard deviation of 1. In order to assure that a given normalized score can indicate the same forest condition in terms of forest intactness across FMUs, we regressed normalized nMDS axis-1 scores with relative abundance of pioneer species (genera) per plot in each FMU. Subsequently, significant differences in the slope and intercept of the regression lines among FMUs were tested with a multi regression analysis. Complete overlap of the regression lines suggests that a given normalized score can indicate the same forest condition across FMUs. Indeed, there were no significant differences in the intercept and slope among the regression lines in Segaliud Lokan, Ratah and Sapulut (Supplementary Materials, Figure S3). The intercept and slope of Roda Mas were exceptionally significantly different from the other FMUs, but the range of the deviation was still within that of the other FMUs. The normalized scores will be used as "nMDS axis-1" in the following analyses.

2.4. Satellite Analysis

2.4.1. Satellite Images and Image Pre-Processing

The estimates of the nMDS axis-1 scores for the entire study area in each FMU were based on Landsat TM (Thematic Mapper) and OLI (Operational Land Imager) imageries. Descriptions of the data set (sensor, path/row and date) of the imageries used are given in the Supplementary Material (Table S1). No significant geolocation errors were observed in the images because the GPS ground positions collected on logging roads corresponded closely with the images.

As pre-processing, the raw digital numbers of each image were converted into top-of-atmosphere radiance. Subsequently, to compensate for atmospheric scattering and absorption effects, an atmospheric correction algorithm based on the Second Simulation of a Satellite Signal in the Solar Spectrum radiative transfer code [24,25] was used to convert top-of-atmosphere radiance into surface reflectance. Finally, the effects of differential illumination due to topography were reduced using the method described by Ekstrand [26]. Shuttle Radar Topography Mission (SRTM) data were used to correct the illumination effects. Then, pixels covered with clouds/shadow were removed using an object-based approach. We created image segments composed of spectrally coherent pixels that were clustered based on homogeneity criteria, and established a threshold to detect cloud/shadow segments. Then, we removed them using the threshold and visual inspection.

Subsequently, the missing data due to cloud cover in Sapulut, Roda Mas, and Ratah were filled in using the cloud-free areas of temporally adjoining data. Where satellite imagery is subjected to high incidence of clouds or haze, which is often the case in tropical rain forests, the mosaicking of cloud-free parts of temporally adjoining data may be the best option for deriving cloud-free coverage [27]. In our study, in the mosaicking procedure, calibrated cloud-free parts of adjoining secondary images were embedded in the missing parts of the base image. To calibrate secondary images to the base image, we used a linear regression model [28–34]. The basis for the model was the set of co-located, mutually clear pixels from each base- and secondary-scene pair. Before establishing regression models, we excluded pixels with marked spectral changes, which possibly occurred with land-cover changes, using a change-detection analysis with a normalized difference vegetation index (NDVI) [35]. Subsequently, pixel-level regression models were established using a robust regression. A regression model for Landsat-TM/OLI images has the following general form:

$$y_{base\,i} = f\,(x_{sec\,1}, x_{sec\,2}, x_{sec\,3}, x_{sec\,4}, x_{sec\,5}, x_{sec\,6}, x_{sec\,7}),$$

where $y_{base\,i}$ is the reflectance value of the pixel in the base image for the ith band to be predicted, and $x_{sec\,1}$ is the reflectance value of band 1 of the co-located pixel in the secondary images, $x_{sec\,2}$ is the reflectance value of band 2, and so on. Missing data in each pixel in the base images were replaced by the calibrated pixels of secondary images using the regression models. Landsat-OLI (Jun/19/2013), Landsat-OLI (May/31/2015), and Landsat-TM (Feb/10/2010) were used as base images in Sapulut, Roda Mas, and Ratah, respectively. eCognition Developer 8.7 was used to create image segments, and ERDAS Imagine ver.11.0 and ArcGIS 9.3.1 were used for pre-processing.

2.4.2. Extrapolation of nMDS Axis-1 Scores Based on Landsat Data

To estimate the nMDS axis-1 scores of the entire study area of each FMU, we established multiple linear regression models between the nMDS axis-1 scores of the inventoried plots and corresponding metrics of the Landsat imagery (i.e., average value of Landsat metrics within the 20 m-radius from the center of the plots) in each FMU. Then, the nMDS axis-1 scores were extrapolated to the entire area of each FMU based on the model. The inventory plots that were unavailable because of cloud cover and could not be replaced in the above cloud-correction procedure were eliminated in this procedure. We developed one independent model for each FMU rather than developing a single model for all

FMUs combined by combining all plots because it was necessary to take regional floristic variation into account.

The normalized scores of nMDS axis-1 were considered response variables, and the Landsat metrics were considered independent variables. We used the following Landsat metrics: reflectance value of each band (Band1$_{TM/OLI}$, Band2$_{TM/OLI}$, Band3$_{TM/OLI}$, Band4$_{TM/OLI}$, Band5$_{TM/OLI}$, Band6$_{OLI}$, Band7$_{TM/OLI}$), normalized difference vegetation index (NDVI) [35], normalized difference water index (NDWI) [36,37], normalized difference soil index (NDSI) [38], and enhanced vegetation index (EVI) [39]. The mean values of the Landsat metrics within a 20 m radius from the center of the plots were used as the metrics corresponding to each plot. The indices were calculated from the following equations:

$$NDVI = (Band4_{TM}(5_{OLI}) - Band3_{TM}(4_{OLI})) \ / \ (Band4_{TM}(5_{OLI}) + Band3_{TM}(4_{OLI})),$$

$$NDWI = (Band3_{TM}(4_{OLI}) - Band5_{TM}(6_{OLI})) \ / \ (Band3_{TM}(4_{OLI}) + Band5_{TM}(6_{OLI})),$$

$$NDSI = (Band5_{TM}(6_{OLI}) - Band4_{TM}(5_{OLI})) \ / \ (Band5_{TM}(6_{OLI}) + Band4_{TM}(5_{OLI})),$$

$$EVI = 2.5*(Band4_{TM}(5_{OLI}) - Band3_{TM}(4_{OLI})) \ / \ (Band4_{TM}(5_{OLI}) + 6*Band3_{TM}(4_{OLI}) - 7.5*Band1_{TM}(2_{OLI}) + 1).$$

In addition to the above metrics, the coefficient of variation (CV), standard deviation (SD), and textures of the gray level co-occurrence matrix (GLCM) [40] were used as proxies for spectral heterogeneity because degradation of forest canopies might affect the heterogeneity of the spectral pattern. The CV, SD, and textures of the GLCM were calculated using a 3×3 pixel window based on the reflectance values and each of the indices. The GLCM is a tabulation of how often different combinations of gray levels occur at a specified distance and orientation in an image object [41]. The homogeneity, contrast, angular second moment, entropy, dissimilarity, correlation, mean, and standard deviation were calculated as the indices of the textures. A total of 120 and 132 metrics were generated based on Landsat TM and OLI, respectively. The independent variables of the regression models were chosen using a stepwise selection from all of the metrics to avoid multi-collinearity among the independent variables. The values of the variance inflation factor (VIF) of selected independent variables were all less than 10 except for Roda Mas, which indicated that there was no multicollinearity (see Supplementary Materials, Table S2). We removed the areas above 600 m elevation from the analysis, because the potential natural vegetation above 600 m differed from that of lowland natural forest [42,43]. eCognition developer 8.7 was used to calculate texture, R ver 3.20 was used to establish the models, and ERDAS Imagine ver.11.0 and ArcGIS 9.3.1 were used in the other procedures.

2.5. Validation of nMDS Axis-1 Model

We used a cross-validation approach [20,44] to assess the accuracy of the models. Four-fifths of all plots were randomly selected in each FMU to construct the nMDS axis-1 models. Based on these models, we estimated the nMDS axis-1 scores of the remaining one-fifth of the plots. Then, we tested the correlation between the estimated nMDS axis-1 scores and the field-measured nMDS axis-1 scores. This step was reiterated 1,000 times in each FMU to derive the 95% confidence interval of the correlation coefficient. The statistical tests were conducted using R ver. 3.20.

2.6. Comparison of Canopy Conditions Based on nMDS Axis-1 Scores among FMUs

To assess whether our method could reliably diagnose the magnitude of forest intactness, we calculated the histogram of relative frequency and mean value of the nMDS axis-1 scores for each FMU and compared them based on their management types. To compare the histograms, we chose four representative FMUs: Roda Mas (FMU practicing reduced-impact logging in relatively intact forests); Deramakot (FMU practicing reduced-impact logging in previously conventionally logged forests); Segaliud Lokan (FMU with a longer history of conventional logging); and Sapulut (FMU including industrial tree plantations). The histograms were represented as probability density functions, which indicate relative frequency of nMDS axis-1 scores in each FMU.

The histogram of the nMDS axis-1 scores of each FMU was used as an index of the spatial variability of the magnitude of forest intactness, and the mean value was used as the average condition in each FMU. To compare the histograms of the nMDS axis-1 scores among FMUs, a regression model based on all plots (approximately 50) was extrapolated to the entire area of the FMU, from which the histogram was derived. To compare the differences in the mean nMDS axis-1 scores among the six FMUs, four-fifths of all plots were randomly selected to construct nMDS axis-1 models in each FMU. Based on these models, we estimated the mean nMDS axis-1 scores of the entire area of each FMU. This step was iteratively done 1000 times to derive the mean value and 95% confidence intervals of the mean nMDS axis-1 scores. Then, the relationships of the derived histograms and means of the nMDS axis-1 scores with the type of forest management were examined.

3. Results

3.1. Forest-Intactness Maps

The nMDS axis-1 scores based on the stepwise selection were mainly explained by the short-wave infrared (SWIR) reflectance, textures, and SD (Table 2). The correlation coefficient scores between the predicted and observed nMDS axis-1 scores ranged from 0.604 to 0.745 (Figure 2). The derived forest-intactness maps of all FMUs are shown in Figure 3. Each pixel in the maps contains an nMDS axis-1 score, and the color gradation from blue to red indicates a gradient of nMDS axis-1 scores from high to low values as a proxy for the canopy tree composition at a genus level. The maps showed conspicuous blue areas corresponding to the tree-community composition of an intact forest as well as conspicuous red areas corresponding to the tree-community composition of the least intact forest. Intact forests and least-intact forests were indicated as the two extremes of a continuum of forest degradation in our method.

Table 2. Description of established multivariate regression models based on all plots.

	R^2	Coefficient	SE	T-value	Pr (>|t|)
SegaliudLokan-Deramakot-Tangkulap (N = 47)					
$B7_{TM}$		−0.38989	2.80E-04	3.69863	<0.001
GLCM_mean_NDVI		−0.25902	1.20E-02	2.23543	3.10E-02
GLCM_correlation_$B3_{TM}$	0.642	−0.32586	3.80E-01	3.3816	1.60E-03
GLCM_mean_$B1_{TM}$		0.30875	9.40E-03	2.76426	8.50E-03
GLCM_homogenity_$B4_{TM}$		0.1891	3.70E+00	2.03608	4.80E-02
Sapulut (N = 45)					
$B6_{OLI}$		−0.46159	2.40E-05	3.7795	<0.001
GLCM_contrast_$B3_{OLI}$		0.37483	3.10E-05	3.8679	<0.001
$B5_{OLI}$	0.604	−0.36385	1.20E-05	3.0253	4.30E-03
GLCM_mean_$B6_{OLI}$		−0.286	8.20E-03	2.8178	7.50E-03
Roda Mas (N = 45)					
$B6_{OLI}$		−0.86698	7.50E-05	4.9699	<0.001
SD_NDSI		−0.75056	1.20E-04	4.5416	<0.001
SD_NDWI	0.745	−2.12925	2.20E-04	4.6781	<0.001
SD_NDVI		2.00116	4.50E-04	3.373	1.70E-03
$B4_{OLI}$		0.56146	4.70E-04	2.3481	2.40E-02
Ratah (N = 64)					
$B7_{TM}$		−1.42357	1.30E-03	7.02905	<0.001
$B3_{TM}$	0.615	0.75859	2.80E-03	3.74366	<0.001
GLCM_dissimilarity_$B4_{TM}$		0.25656	9.80E-03	3.2506	1.90E-03

Note: N, number of plots used for each model; R^2, adjusted R-squared value; Coefficient, standardized partial regression coefficient; SE, standard error; SD, standard deviation; GLCM, grey level co-occurrence matrix; NDVI, normalized difference vegetation index; NDSI, normalized difference soil index; NDWI, normalized difference water index.

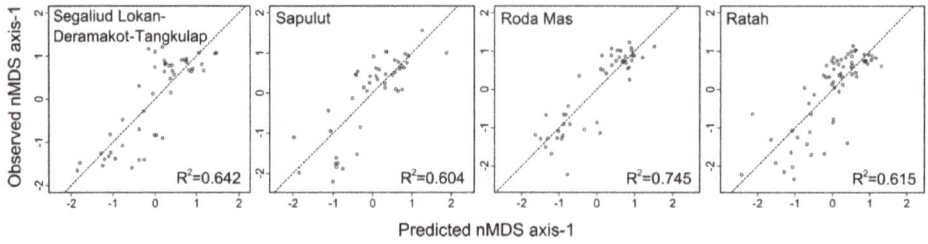

Figure 2. Model fits for all available inventory plots in each FMU. Scatter plots show relationships between predicted and observed scores.

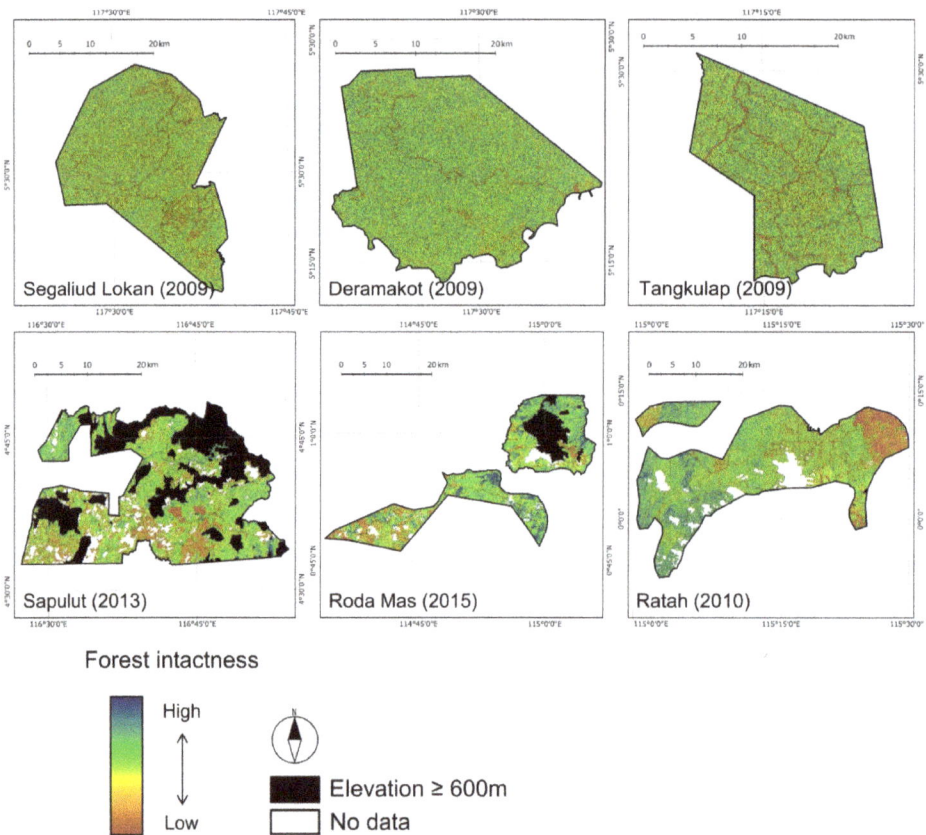

Forest intactness

Figure 3. Tree-community composition maps of all FMUs. A color gradation from **blue** to **red** indicates a gradient of normalized nMDS axis-1 scores from high to low values as a proxy for canopy tree composition at a genus level.

Several FMUs had areas of intact forests (e.g., along the left side in Ratah and upper-right side in Roda Mas). A few of the FMUs had conspicuous red-yellow areas, which corresponded to highly degraded areas. The most conspicuous red-yellow area was the eastern part of Ratah, which corresponded to the land after forest fire in 1998. The red-yellow area at western part of Roda Mas was allocated to local people who practice slash and burn agriculture. Mosaics of red areas occurred in Sapulut, which corresponded to industrial tree plantations. The forest intactness inside the Deramakot was uniformly green, reflecting the history of reduced-impact logging and longer cutting cycle (i.e., 40 years). In contrast, Segaliud Lokan where reduced-impact logging had been implemented after 2002 and Tangkulap where all logging activities had remained suspended after 2002 had sparse yellow-red areas inside. This indicated that the negative effect of high-impact conventional logging continued until 2002 still remained in these two FMUs.

3.2. Validation of Community-Composition (nMDS Axis-1) Models

The cross validation indicated that the accuracy of the model estimates varied among FMUs but was generally high. The means (95% confidence interval, CI) of the iteratively calculated correlation coefficients between the estimated nMDS axis-1 scores and actual observed scores (adjusted R^2 values) were 0.57 (CI 0.10–0.87), 0.54 (CI 0.10–0.87), 0.69 (CI 0.37–0.91), and 0.56 (CI 0.12–0.84) for Segaliud Lokan-Deramakot-Tangkulap, Sapulut, Roda Mas, and Ratah, respectively.

3.3. Comparison of Histogram/Mean nMDS Axis-1 Scores among FMUs

The histograms of the nMDS axis-1 scores in six FMUs are shown in Figure 4a. The histograms varied significantly among the FMUs in terms of mode and pattern; the ranges were the same because the nMDS axis-1 scores were standardized across the FMUs. The mode of Roda Mas had the highest axis-1 score among the FMUs, indicating that relatively intact forests occur over a disproportionately greater area in Roda Mas. At the same time, the density of low nMDS axis-1 scores (nMDS axis-1 score = −1) was also nominal in Roda Mas, indicating the occurrence of current logging activities with reduced-impact logging techniques. The mode in Deramakot, the model site of sustainable forest management in Sabah, had the second highest score among the FMUs. The nMDS axis-1 scores of the modes among four representative FMUs were decreased in the order reflecting their logging intensity, history or management (Figure 4b).

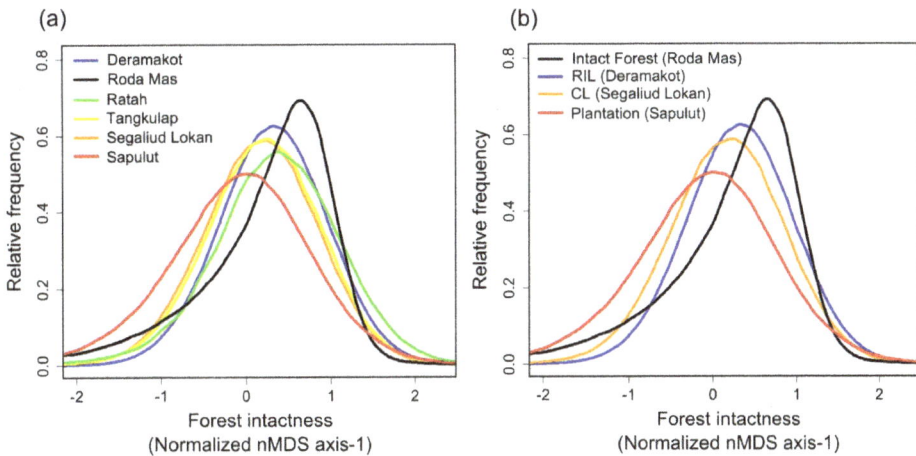

Figure 4. Histograms of normalized nMDS (non metric multidimensional scaling) axis-1 scores (**a**) in all FMUs, and (**b**) in four selected FMUs. (RIL, reduced-impact logging; CL, conventional logging).

The mean values and 95% confidence intervals of the nMDS axis-1 scores are shown in Figure 5, in which the six FMUs are arranged in increasing order of their mean nMDS axis-1 score. Deramakot, which has been designated as a model of sustainable forest management by the Sabah Government, recorded the highest mean score, 0.318 (CI 0.197–0.432). The second-highest score was 0.289 (CI 0.219–0.365) in Roda Mas. The scores were 0.220 (CI 0.151–0.293) in Ratah. Tangkulap and

Segaliud Lokan, where high-impact conventional logging continued until 2002, had relatively low scores of 0.192 (CI 0.094–0.290) and 0.163 (CI 0.063–0.262). Sapulut, where industrial tree plantations occur in approximately 35% of the area, recorded the lowest score of -0.082 (CI -0.184–0.027). Overall, the mean score of Sapulut (-0.082) was significantly lower than that of the other FMUs, while the other FMUs did not significantly differ from one another. There was well-defined variation in the mean score (0.192–0.318) among the FMUs that were certified by the FSC (Figure 5), although the 95% confidence intervals broadly overlapped with each other among the FMUs.

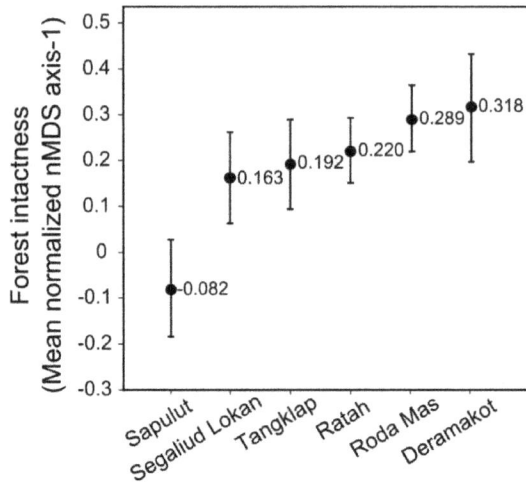

Figure 5. Mean values of normalized nMDS axis-1 scores in all FMUs.

4. Discussion

To address the increasing need for a practical method of assessing biodiversity that is spatially and temporally sensitive and robust, we developed an algorithm to map forest intactness based on tree-community composition. As a proxy for tree-community composition, we used the nMDS axis-1 score, a robust and sensitive indicator of forest intactness in Bornean natural production forests [12]. The maps based on the algorithm demonstrated a significant variation in composition of tree genera both among and within FMUs.

To date, several remote sensing methods have been developed to map tree-community composition using imaging spectroscopy [45–49]. Where imaging spectroscopy (also known as hyperspectral imagery) can characterize variation in tree compositional assemblages, multispectral imageries (e.g., Landsat) have been considered to lack the spectral resolution to effectively demonstrate compositional variation [50,51]. However, our study shows that multispectral remote-sensing data of Landsat images are potentially useful for characterizing the variation in composition of tree genera caused by logging and forest degradation. The concept of mapping compositional gradients using remote sensing is based on the assumption that the compositional gradients of vegetation are related to variation in canopy traits such as reflectance spectra, foliar chemistry, morphological traits, canopy form, and canopy structure [48]. Clearly, Landsat imagery may not effectively resolve subtle intra-guild trait variation (e.g., trait variation within climax guild) because of the low spatial/spectral resolution; however, Landsat imagery has been proven to be capable of detecting inter-guild trait differences (i.e., trait differences between pioneer and climax guilds) [52,53] as shown by our algorithm. The reason why the proxy for community composition (nMDS axis-1 score) used in our study is able to reflect canopy intactness is that the nMDS axis-1 scores can reflect the interaction of linearly increasing pioneer guild and linearly decreasing climax guild with increasing logging intensity [12]. Thus, the selected independent variables in our models and Landsat imageries reflect the interactions of these guilds. The variables selected by stepwise selection were principally related to SWIR reflectance and spectral heterogeneity information (i.e., texture and SD). SWIR reflectance is significantly related to stand age, height, volume, and biomass in tropical secondary forest [54]. In contrast, spectral heterogeneity information can be considered a proxy for species diversity [55]. These patterns were also apparent in our study. SWIR reflectance was a proxy for tree growth (i.e., biomass) after logging, while texture and SD were related to the heterogeneity of the two major regeneration guilds. We suggest that SWIR reflectance and spectral heterogeneity information can reflect the interactions of the two guilds, and our algorithm is thus suitable for the characterization of the variation in composition of tree genera caused by logging disturbances.

It should also be noted that the selected independent variables and coefficients varied among models. Our study area covered the entire area of Borneo, spanning a large geographic area, and included significant floristic differences among the tropical rain forests [56]. Therefore, the selected independent variables would be expected to vary among FMUs, reflecting variation in canopy structure and foliar traits.

The mean adjusted R^2 values (0.54–0.69) based on the cross-validation approach showed close correlation in all FMUs. The mean adjusted R^2 values were high despite the use of moderate-resolution Landsat data and the semi-qualitative nature of our metrics (i.e., tree-community composition), which were based on the mixing ratio of the canopy genera. However, the range of adjusted R^2 values (0.1–0.9) at the 95% confidence interval across all FMUs was fairly wide; this probably indicates that the number of plots used for cross validation was too small. Depending on the combination of plots used for developing the models, the correlation between the predicted and observed nMDS axis-1 scores became very low. Because of the wide variation in the R^2 values at the 95% confidence interval, it was difficult to discriminate the FMUs from one another, despite the fact that the mean nMDS axis-1 scores per se differed considerably among FMUs (Figure 5). Overall, we were able to differentiate only one FMU with the lowest mean nMDS axis-1 score (Sapulut) from the other FMUs (Figure 5) in our analysis. However, our method is clearly useful for differentiating between intact and highly disturbed forests within each FMU when all 50 plots are used to develop the best model. It can identify the location and extent of intact forests as areas of potentially high conservation value, or those of least-intact forests where mitigation measures are required. In combination with a biomass estimate, our method is also useful for identifying the location and extent of less-intact forests with high biomass, indicative of plantations. It thus offers a potential tool for verifying compliance with environmental safeguards in REDD+, in which a primary concern is the conversion of natural forests into plantation.

The applicability of our method was verified over a broad geographic range (0°7′S–5°20′N, 114°25′–115°30′E) spanning the most area of Borneo. The principle of using tree-community composition may also be applicable to other forests outside Borneo because the occurrence of the two major regeneration guilds (i.e., the pioneer and climax guilds) is a common biological phenomenon; however, this needs to be tested in the future. One of the advantages of this method is the broad coverage and low cost based on Landsat imagery. Moreover, our analysis of community composition is based on canopy-tree genera only. As Imai et al. [12] pointed out, identification of canopy trees at the generic level can reduce the cost and time of identification by at least 60% compared with the conventional routine survey at species level. The spatial and temporal availability of Landsat imagery is superior to other remote sensing data, and therefore the Landsat-based tree-community composition map is suitable for regional and global biodiversity monitoring. These advantages are essential for a global biodiversity assessment of the Aichi Biodiversity Targets and REDD+ biodiversity safeguards. We propose a practical method combining count-plot surveys on the ground with Landsat remote sensing for large-scale forest biodiversity/ecosystem assessments of both the Aichi targets and REDD+ biodiversity safeguards in natural forests in the tropics.

Supplementary Materials: The following are available online at http://www.mdpi.com/2073-445X/5/4/45/s1, **Table S1.** Description of the Landsat imagery data set used in this study, including sensor, path / row and acquisition date. **Table S2.** Variance inflation factor (VIF) of the selected variable in the established multivariate regression models. **Figure S1.** The effects of distance on the difference in the nMDS axis-1 scores of paired vegetation plots. See the main text for the details of the prediction model of nMDS axis-1 for each FMU. Positive relationships were observed for Ratah, potentially indicating the presence of spatial autocorrelations, but adjusted R^2 scores were extremely low in all FMUs: SegaliudLokan-Deramakot-Tangkulap, $R^2 = -0.00$, $P > 0.05$, n.s.; Sapulut, $R^2 = 0.00$, $P > 0.05$, n.s.; Roda Mas, $R^2 = 0.00$, $P > 0.05$, n.s.; Ratah, $R^2 = 0.03$, $P < 0.0001$. n.s. denotes non-significance. Moreover, positive relationships may simply represent the consolidated occurrence of intact forests or disturbed forests. **Figure S2.** The effects of distance on the difference in the residuals of nMDS axis-1 (i.e., observed nMDS axis-1 − predicted nMDS axis-1) of paired vegetation plots. See the main text for the details of the prediction model of nMDS axis-1 for each FMU. Positive relationships were observed for two FMUs (Sapulut and Roda Mas), potentially indicating the presence of spatial autocorrelations, but adjusted R^2 scores were extremely low in all FMUs: SegaliudLokan-Deramakot-Tangkulap, $R^2 = -0.00$, $P > 0.05$, n.s.; Sapulut, $R^2 = 0.01$, $P < 0.001$; Roda Mas, $R^2 = 0.01$, $P < 0.0001$; Ratah, $R^2 = 0.00$, $P > 0.05$, n.s. denotes non-significance. Moreover, positive relationships may simply represent the consolidated occurrence of intact forests or disturbed forests. **Figure S3.** The relationships between the normalized nMDS axis-1 scores and relative abundance of pioneer species (genera) in each FMU. The adjusted R^2 scores were significantly high in all FMUs: SegaliudLokan-Deramakot-Tangkulap, $R^2 = 0.72$, $P < 0.001$; Sapulut, $R^2 = 0.90$, $P < 0.001$; Roda Mas, $R^2 = 0.95$, $P < 0.001$; Ratah, $R^2 = 0.82$, $P < 0.001$. There were no significant differences in the intercept and slope of regression lines among Segaliud Lokan (including Deramakot and Tangkulap), Ratah and Sapulut. The intercept and slope of Roda Mas were exceptionally significantly different from the other FMUs (intercept, $P < 0.05$; slope, $P < 0.001$), but the range of the deviation was still within that of the other FMUs.

Acknowledgments: We acknowledge the kind assistance from WWF (World Wide Fund for Nature) Indonesia, WWF Japan, the Agriculture and Forestry Services of the District of Mahakam Ulu (East Kalimantan), Sabah Forestry Department and Sabah Forest Research Centre, Sarawak Forest Department, Botanical Research Center, Sarawak Forestry Corporation, the KTS Plantation Sdn. Bhd., the Sapulut Forest Development Sdn. Bhd., Zedtee Sdn. Bhd., the PT (Perseroan Terbatas) Roda Mas Timber and the PT Ratah Timber. We are thankful to Miun Poster, Joel Bin Dawat and Sutrisno for tree species identification; Itong Sarjuni and the staff of the FMUs for assisting field inventory; and Dodit Agus Riyono, and Mutai Hashimoto for generous support for every aspect. This study was supported by the Global Environment Research Fund 1-1403 of the Ministry of the Environment, Japan, to Kanehiro Kitayama.

Author Contributions: K.K. conceived and designed the research; S.F. invented the algorithm; S.F., R.A., A.T., N.I. and H.S. collected and analyzed the data; and all authors wrote the paper.

Conflicts of Interest: The authors declare that there is no conflict of interests regarding the publication of this paper. The founding sponsors had no role in the design of the study; in the collection, analyses, or interpretation of data; in the writing of the manuscript, and in the decision to publish the results.

References

1. Hansen, M.C.; Potapov, P.V.; Moore, R.; Hancher, M.; Turubanova, S.; Tyukavina, A.; Thau, D.; Stehman, S.; Goetz, S.; Loveland, T. High-resolution global maps of 21st-century forest cover change. *Science* **2013**, *342*, 850–853. [CrossRef] [PubMed]

2. CBD (Convention on Biological Diversity). CBD quick guides to the Aichi Biodiversity Targets: 2014. Available online: https://www.cbd.int/nbsap/training/quick-guides/ (accessed on 14 September 2015).

3. GEO BON Office. *Adequacy of Biodiversity Observation Systems to Support the CBD 2020 Targets, A Report Prepared by the Group on Earth Observations Biodiversity Observation Network (GEO BON), for the Convention on Biological Diversity*; GEO BON Office: Pretoria, South Africa, 2011.

4. Pereira, H.M.; Ferrier, S.; Walters, M.; Geller, G.N.; Jongman, R.; Scholes, R.J.; Bruford, M.W.; Brummitt, N.; Butchart, S.; Cardoso, A. Essential biodiversity variables. *Science* **2013**, *339*, 277–278. [CrossRef] [PubMed]

5. Miles, L.; Kapos, V. Reducing greenhouse gas emissions from deforestation and forest degradation: Global land-use implications. *Science* **2008**, *320*, 1454–1455. [CrossRef] [PubMed]

6. Paoli, G.D.; Wells, P.L.; Meijaard, E.; Struebig, M.J.; Marshall, A.J.; Obidzinski, K.; Tan, A.; Rafiastanto, A.; Yaap, B.; Slik, J.F. Biodiversity conservation in the REDD. *Carbon Balance Manag.* **2010**, *5*, 7. [CrossRef] [PubMed]

7. CCBA. *Climate, Community & Biodiversity Project Design Standards*, 2nd ed.; CCBA: Arlington, VA, USA, 2008.

8. Gardner, T.A.; Burgess, N.D.; Aguilar-Amuchastegui, N.; Barlow, J.; Berenguer, E.; Clements, T.; Danielsen, F.; Ferreira, J.; Foden, W.; Kapos, V. A framework for integrating biodiversity concerns into national REDD+ programmes. *Biol. Conserv.* **2012**, *154*, 61–71. [CrossRef]

9. GCS. *Global Conservation Standard Version 1.2*; Global Conservation Standard e.V.: Offenburg, Germany, 2011.

10. FSC (Forest Stewardship Council). Briefing Paper: Preliminary Outreach to FSC Membership in Preparation for the Development of FSC International Generic Indicators. 2012. Available online: http://igi.fsc.org/download.fsc-generic-indicators-outreach-briefing.28.pdf (accessed on 28 August 2015).

11. Su, J.C.; Debinski, D.M.; Jakubauskas, M.E.; Kindscher, K. Beyond species richness: Community similarity as a measure of cross-taxon congruence for coarse-filter conservation. *Conserv. Biol.* **2004**, *18*, 167–173. [CrossRef]

12. Imai, N.; Tanaka, A.; Samejima, H.; Sugau, J.B.; Pereira, J.T.; Titin, J.; Kurniawan, Y.; Kitayama, K. Tree community composition as an indicator in biodiversity monitoring of REDD+. *Forest Ecol. Manage.* **2014**, *313*, 169–179. [CrossRef]

13. Kitayama, K. (Ed.) *Co-Benefits of Sustainable Forestry: Ecological Studies of a Certified Bornean Rain Forest*; Springer Science & Business Media: Tokyo, Japan, 2012.

14. Sabah Forestry Department. *Forest Management Plan 2: Deramakot Forest Reserve, Forest Management Unit No. 19*; Sabah Forestry Department: Sandakan, Malaysia, 2005.

15. Applegate, G.; Kartawinata, K.; Klassen, A. *Reduced Impact Logging Guidelines for Indonesia*; CIFOR: Bogor, Indonesia, 2001.

16. Sabah Forestry Department. *RIL Operation Guide Book*, 3rd ed.; Sabah Forestry Department: Sandakan, Malaysia, 2009.

17. Lagan, P.; Mannan, S.; Matsubayashi, H. Sustainable use of tropical forests by reduced-impact logging in Deramakot Forest Reserve, Sabah, Malaysia. *Ecol. Res.* **2007**, *22*, 414–421. [CrossRef]

18. Pinard, M.A.; Putz, F.E. Retaining forest biomass by reducing logging damage. *Biotropica* **1996**, *28*, 278–295. [CrossRef]

19. Putz, F.E.; Zuidema, P.A.; Pinard, M.A.; Boot, R.G.; Sayer, J.A.; Sheil, D.; Sist, P.; Vanclay, J.K. Improved tropical forest management for carbon retention. *PLoS Biol.* **2008**, *6*, e166. [CrossRef] [PubMed]

20. Langner, A.; Samejima, H.; Ong, R.C.; Titin, J.; Kitayama, K. Integration of carbon conservation into sustainable forest management using high resolution satellite imagery: A case study in Sabah, Malaysian Borneo. *Int. J. Appl. Earth Obs. Geoinf.* **2012**, *18*, 305–312. [CrossRef]

21. Penman, J.; Gytarsky, M.; Hiraishi, T.; Krug, T.; Kruger, D.; Pipatti, R.; Buendia, L.; Miwa, K.; Ngara, T.; Tanabe, K. *Definitions and Methodological Options to Inventory Emissions from Direct Human-Induced Degradation of Forests and Devegetation of Other Vegetation Types*; IPCC National Greenhouse Gas Inventories Programme-Technical Support Unit: Hayama Kanagawa, Japan, 2003; p. 32. Available online: http://www.ipcc-nggip.iges.or.jp (accessed on 21 February 2016).

22. Chao, A.; Chazdon, R.L.; Colwell, R.K.; Shen, T.J. A new statistical approach for assessing similarity of species composition with incidence and abundance data. *Ecol. Lett.* **2005**, *8*, 148–159. [CrossRef]

23. Oksanen, J.; Blanchet, F.G.; Kindt, R.; Legendre, P.; Minchin, P.R.; O'Hara, R.; Simpson, G.L.; Solymos, P.; Stevens, M.H.H.; Wagner, H. *Package 'vegan'. Community Ecology Package*, version 2.0-9. 2013.

24. Kotchenova, S.Y.; Vermote, E.F.; Matarrese, R.; Klemm, F.J., Jr. Validation of a vector version of the 6S radiative transfer code for atmospheric correction of satellite data. Part I: Path radiance. *Appl. Opt.* **2006**, *45*, 6762–6774. [CrossRef] [PubMed]

25. Vermote, E.F.; Tanré, D.; Deuze, J.L.; Herman, M.; Morcette, J.-J. Second simulation of the satellite signal in the solar spectrum, 6S: An overview. *IEEE Trans. Geosci. Remote Sens.* **1997**, *35*, 675–686. [CrossRef]

26. Ekstrand, S. Landsat TM-based forest damage assessment: Correction for topographic effects. *Photogramm. Eng. Remote Sens.* **1996**, *62*, 151–162.

27. Helmer, E.; Ruefenacht, B. Cloud-free satellite image mosaics with regression trees and histogram matching. *Photogramm. Eng. Remote Sens.* **2005**, *71*, 1079–1089. [CrossRef]

28. Schott, J.R.; Salvaggio, C.; Volchok, W.J. Radiometric scene normalization using pseudoinvariant features. *Remote Sens. Environ.* **1988**, *26*, 1IN115–1416. [CrossRef]

29. Vogelmann, J.E. Detection of forest change in the Green Mountains of Vermont using multispectral scanner data. *Int. J. Remote Sens.* **1988**, *9*, 1187–1200. [CrossRef]

30. Hall, F.G.; Strebel, D.E.; Nickeson, J.E.; Goetz, S.J. Radiometric rectification: Toward a common radiometric response among multidate, multisensor images. *Remote Sens. Environ.* **1991**, *35*, 11–27. [CrossRef]

31. Olsson, H. Regression functions for multitemporal relative calibration of Thematic Mapper data over boreal forest. *Remote Sens. Environ.* **1993**, *46*, 89–102. [CrossRef]
32. Oetter, D.R.; Cohen, W.B.; Berterretche, M.; Maiersperger, T.K.; Kennedy, R.E. Land cover mapping in an agricultural setting using multiseasonal Thematic Mapper data. *Remote Sens. Environ.* **2001**, *76*, 139–155. [CrossRef]
33. Song, C.; Woodcock, C.E.; Seto, K.C.; Lenney, M.P.; Macomber, S.A. Classification and change detection using Landsat TM data: When and how to correct atmospheric effects? *Remote Sens. Environ.* **2001**, *75*, 230–244. [CrossRef]
34. Du, Y.; Teillet, P.M.; Cihlar, J. Radiometric normalization of multitemporal high-resolution satellite images with quality control for land cover change detection. *Remote Sens. Environ.* **2002**, *82*, 123–134. [CrossRef]
35. Rouse, J.; Haas, R.; Schell, J.; Deering, D.; Harlan, J. *Monitoring the Vernal Advancement and Retrogradation of Natural Vegetation*; NASA/GSFC Type III Final Report; NASA/GSFC: Greenbelt, MD, USA, 1974; p. 371.
36. Gao, B.-C. NDWI—A normalized difference water index for remote sensing of vegetation liquid water from space. *Remote Sens. Environ.* **1996**, *58*, 257–266. [CrossRef]
37. McFeeters, S.K. The use of the Normalized Difference Water Index (NDWI) in the delineation of open water features. *Int. J. Remote Sens.* **1996**, *17*, 1425–1432. [CrossRef]
38. Takeuchi, W.; Yasuoka, Y. Development of normalized vegetation, soil and water indices derived from satellite remote sensing data. *J. Jpn Soc. Photogramm. Remote Sens.* **2004**, *43*, 7–19. [CrossRef]
39. Huete, A.; Didan, K.; Miura, T.; Rodriguez, E.P.; Gao, X.; Ferreira, L.G. Overview of the radiometric and biophysical performance of the MODIS vegetation indices. *Remote Sens. Environ.* **2002**, *83*, 195–213. [CrossRef]
40. Haralick, R.M. Statistical image texture analysis. In *Handbook of Pattern Recognition and Image Processing*; Young, T.Y., Fu, K.S., Eds.; Academic Press: Orlando, FL, USA, 1986; Volume 86, pp. 247–279.
41. Baatz, M.; Benz, U.; Dehghani, S.; Heynen, M.; Höltje, A.; Hofmann, P.; Lingenfelder, I.; Mimler, M.; Sohlbach, M.; Weber, M. *eCognition Professional User Guide 4*; Definiens Imaging: Munich, Germany, 2004.
42. Kitayama, K. An altitudinal transect study of the vegetation on Mount Kinabalu, Borneo. *Vegetatio* **1992**, *102*, 149–171. [CrossRef]
43. Aiba, S.-I.; Kitayama, K. Structure, composition and species diversity in an altitude-substrate matrix of rain forest tree communities on Mount Kinabalu, Borneo. *Plant Ecology* **1999**, *140*, 139–157. [CrossRef]
44. Roff, D.A. *Introduction to Computer-Intensive Methods of Data Analysis in Biology*; Cambridge University Press: Cambridge, UK; New York, NY, USA, 2006.
45. Schmidtlein, S.; Sassin, J. Mapping of continuous floristic gradients in grasslands using hyperspectral imagery. *Remote Sens. Environ.* **2004**, *92*, 126–138. [CrossRef]
46. Schmidtlein, S.; Zimmermann, P.; Schüpferling, R.; Weiss, C. Mapping the floristic continuum: Ordination space position estimated from imaging spectroscopy. *J. Veg. Sci.* **2007**, *18*, 131–140. [CrossRef]
47. Feilhauer, H.; Schmidtlein, S. Mapping continuous fields of forest alpha and beta diversity. *Appl. Veg. Sci.* **2009**, *12*, 429–439. [CrossRef]
48. Feilhauer, H.; Faude, U.; Schmidtlein, S. Combining Isomap ordination and imaging spectroscopy to map continuous floristic gradients in a heterogeneous landscape. *Remote Sens. Environ.* **2011**, *115*, 2513–2524. [CrossRef]
49. Gu, H.; Singh, A.; Townsend, P.A. Detection of gradients of forest composition in an urban area using imaging spectroscopy. *Remote Sens. Environ.* **2015**, *167*, 168–180. [CrossRef]
50. Asner, G.P.; Martin, R.E. Airborne spectranomics: Mapping canopy chemical and taxonomic diversity in tropical forests. *Front. Ecol. Environ.* **2009**, *7*, 269–276. [CrossRef]
51. Rocchini, D. Effects of spatial and spectral resolution in estimating ecosystem α-diversity by satellite imagery. *Remote Sens. Environ.* **2007**, *111*, 423–434. [CrossRef]
52. Wittmann, F.; Anhuf, D.; Funk, W.J. Tree species distribution and community structure of central Amazonian várzea forests by remote-sensing techniques. *J. Trop. Ecol.* **2002**, *18*, 805–820. [CrossRef]
53. Tangki, H.; Chappell, N.A. Biomass variation across selectively logged forest within a 225-km^2 region of Borneo and its prediction by Landsat TM. *For. Ecol. Manag.* **2008**, *256*, 1960–1970. [CrossRef]
54. Steininger, M. Satellite estimation of tropical secondary forest above-ground biomass: Data from Brazil and Bolivia. *Int. J. Remote Sens.* **2000**, *21*, 1139–1157. [CrossRef]

55. Rocchini, D.; Balkenhol, N.; Carter, G.A.; Foody, G.M.; Gillespie, T.W.; He, K.S.; Kark, S.; Levin, N.; Lucas, K.; Luoto, M. Remotely sensed spectral heterogeneity as a proxy of species diversity: Recent advances and open challenges. *Ecol. Inform.* **2010**, *5*, 318–329. [CrossRef]
56. Slik, J.; Poulsen, A.; Ashton, P.; Cannon, C.; Eichhorn, K.; Kartawinata, K.; Lanniari, I.; Nagamasu, H.; Nakagawa, M.; Van Nieuwstadt, M. A floristic analysis of the lowland dipterocarp forests of Borneo. *J. Biogeogr.* **2003**, *30*, 1517–1531. [CrossRef]

Article

Accounting for the Drivers that Degrade and Restore Landscape Functions in Australia

Richard Thackway [1,*] and David Freudenberger [2]

[1] School of Geography, Planning and Environmental Management, University of Queensland, Brisbane, QLD 4072, Australia

[2] Fenner School of Environment & Society, Australian National University, Linnaeus Way, Acton, ACT 2601, Australia; david.freudenberger@anu.edu.au

* Correspondence: r.thackway@uq.edu.au; Tel.: +61-426-258-361

Academic Editors: Jeffrey Sayer and Chris Margules
Received: 25 July 2016; Accepted: 4 November 2016; Published: 12 November 2016

Abstract: Assessment and reporting of changes in vegetation condition at site and landscape scales is critical for land managers, policy makers and planers at local, regional and national scales. Land management, reflecting individual and collective values, is used to show historic changes in ecosystem structure, composition and function (regenerative capacity). We address the issue of how the resilience of plant communities changes over time as a result of land management regimes. A systematic framework for assessing changes in resilience based on measurable success criteria and indicators is applied using 10 case studies across the range of Australia's agro-climate regions. A simple graphical report card is produced for each site showing drivers of change and trends relative to a reference state (i.e., natural benchmark). These reports enable decision makers to quickly understand and assimilate complex ecological processes and their effects on landscape degradation, restoration and regeneration. We discuss how this framework assists decision-makers explain and describe pathways of native vegetation that is managed for different outcomes, including maintenance, replacement, removal and recovery at site and landscape levels. The findings provide sound spatial and temporal insights into reconciling agriculture, conservation and other competing land uses.

Keywords: land management; ecosystem structure; composition; function; tracking change; monitoring; reporting; anthropogenic; transformation; plant communities; vegetation

1. Introduction

Landscapes are dynamic through time, and changes can be efficiently tracked by monitoring the removal, replacement, enhancement or restoration of native vegetation cover. Such dynamics are related to social, economic and political drivers, as well as environmental drivers, such as climate variation. Landscapes are often transformed by intentional or inadvertent management practices that alter native vegetation cover, variously fragmenting it into a matrix of altered states [1–3]. Knowledge of how land management practices are used by local communities to modify and replace native vegetation over time, coupled with landscape genesis and climate variables, can be used to generate predictable pathways for native vegetation recovery (resilience) at sites and landscape scales [4]. This information provides a critical understanding essential for facilitating desired landscape scale change.

For this paper, we ambitiously set ourselves the task of describing and understanding circa 250 years of landscape transformation pathways of native ecosystems from case studies in 10 contrasting ago-climatic regions of Australia. We demonstrate and discuss the value of using a systematic and comprehensive chronology of land management practices and their impacts on vegetation structure, composition and function to document and illustrate the interaction of people living in, learning from and adapting to their environment. We note the value of using a repeatable framework to quantify landscape dynamics over centuries to provide a strong understanding of historical legacies and insights into ecosystem resilience and active restoration. We focus on the benefits of systematically applying this framework to provide rigorous and consistent information for land use planners, policy makers and land managers by elucidating the drivers of such landscape changes. We demonstrate that the condition of landscapes at key points in time is an emergent property of economic markets, the history of settlement, environmental constraints, government policies and programs and the impact of individual land manager's practices and values. We discuss our findings in light of the landscape management principles of Sayer and others [5]; particularly multiple scales; clarification of rights and responsibilities; and resilience. We discuss the theoretical and practical underpinning of how land management effects and changes ecosystem resilience over time [4].

In Australia, landscape transformation is usually assessed relative to a pre-1750 reference state, i.e., at the time of European settlement, or the start of the Industrial Revolution [6]. This date is prior to European settlement, in 1788, and thus reflects the pre-European land-use status, which although not without substantial human effects on the landscape, had been relatively stable for many tens of thousands of years of Aboriginal land use [7,8].

We use the Vegetation Assets States and Transition (VAST) framework [3,9], as a site to landscape assessment tool to critically appraise the relevance of scientific studies, reports and historical knowledge of on-ground practice to document and account for changes in vegetation structure, composition and function. We aimed to demonstrate that the VAST methodology can contribute to a collective learning spiral [10] that may facilitate individuals, communities and government agencies to better understand ecosystem resilience and effectively improve landscape function to deliver a variety of desired ecosystem services [11].

Our premise of applying the VAST methodology is that it can result in improved understanding and management of the key functional, structural and compositional components of the ecosystem at a site or landscape scale relative to its reference state. We argue that within the limits imposed by regional climate and microclimates, the recovery of native ecosystems largely depends on restoring landscape function (e.g., soil structure, hydrology and nutrient cycling), vegetation structure and compositional diversity. The natural disturbance regime, e.g., fire, droughts and floods, must also be documented and understood to effectively manage landscapes sustainably.

We demonstrate that site-based assessments of the effects that local land managers have on indicators of vegetation condition can be up-scaled to generate a whole of landscape perspective. Examples of how the VAST framework has been used to generate national assessments of change in the extent and condition of ecosystems include Australia [2] and Israel [12].

2. Method

2.1. Selecting Case Study Sites to Represent Australia's Agro-Climatic Regions

Ten sites were selected, with one site being assigned to each of 10 agro-climatic regions we have delineated across Australia [13,14] (Figure 1; refer also to Tables S1 and S2 in the Supplementary Material).

At a national scale, there is broad correspondence between agro-climatic regions [13,14] and agro-ecological regions [15], regarding associated patterns of land use and management and both native and human-managed vegetation.

At the international level, agro-climatic categories correspond reasonably well to the primary and secondary Köppen divisions of the 10 global agro-climatic categories [16], but at the finer landscape scale, there are obvious differences in detail [14]. For this reason, we selected a modified form of Australia's 10 agro-climatic regions [13] to provide a broad stratification to assess the extent to which local individuals and communities have been transforming landscapes over the past 250 years.

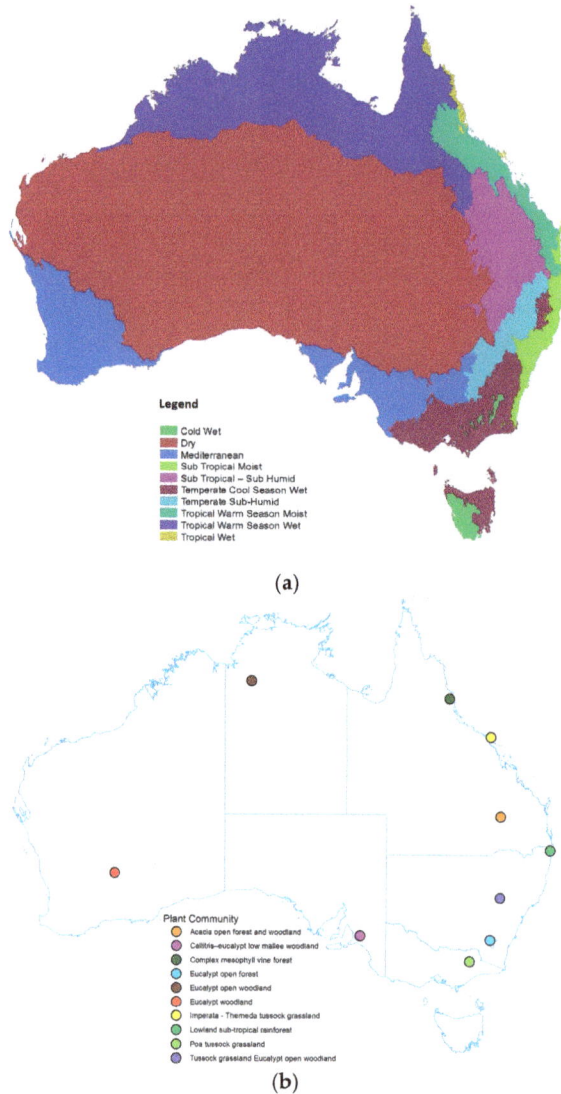

Legend

- Cold Wet
- Dry
- Mediterranean
- Sub Tropical Moist
- Sub Tropical – Sub Humid
- Temperate Cool Season Wet
- Temperate Sub-Humid
- Tropical Warm Season Moist
- Tropical Warm Season Wet
- Tropical Wet

(a)

Plant Community

- Acacia open forest and woodland
- Callitris–eucalypt low mallee woodland
- Complex mesophyll vine forest
- Eucalypt open forest
- Eucalypt open woodland
- Eucalypt woodland
- Imperata - Themeda tussock grassland
- Lowland sub-tropical rainforest
- Poa tussock grassland
- Tussock grassland Eucalypt open woodland

(b)

Figure 1. (**a**) Extent of the 10 agro-climatic regions; and (**b**) location of the 10 sites and the associated plant community type (refer to Tables S1 and S2 in the Supplementary Material).

We acknowledge that plant community responses will differ between different agro-climatic regions and that no two plant community types will have the same responses to different land management practices (i.e., resilience of ecosystems) over time. It follows that the categories of indicators (structure, composition and function) [17] can be the same for all landscapes, while the actual indicators are tied to place.

The 10 agro-climatic regions are described using characteristics of climate, pre-European native vegetation and current land use for each region (refer to Table S2 in the Supplementary Material). Each of the 10 sites is considered typical of the soil-landscape associations, plant communities, land use and management histories of each agro-climatic region.

We merged the agro-climatic regions [13,14] and the agro-ecological regions [15] to derive the regionalisation shown in Figure 1a. We adopted the labels and most of the boundaries of the agro-ecological regions [13] and integrated these with the ecological descriptions of climate, topography, vegetation and land use of the agro-ecological regions [15]. We modified the boundary of the dry and tropical warm season wet agro-climate regions [13] to correspond to the southern extent of the Australian tropical savannas [18]. This boundary adjustment was done acknowledging that Region H was described as an ecotone between tropical warm season wet and the dry continental region [14]. We delimited a revised northern boundary of the dry agro-climate region by using the northern extent of Region G [14]. An added reason for modifying the boundary of the dry and tropical warm season wet agro-climate regions is because, since the 1970s, the rainfall pattern has changed across the tropical savannas. Generally, more rain has been received in the summer period, and the wet season has extended beyond the previously-accepted patterns of summer dominant rains into the autumn period. This change has coincided with a general increase in the cover and density of woody tree cover across the tropical savannas [18,19].

2.2. A Framework for Assessing Change

The VAST framework [2,9] was used as a tool for consistently and repeatedly assessing the effects that land management practices have on the structure, composition and function of plant communities over time. The hierarchical framework of VAST-2 captures the key stages of the degradation and recovery of ecosystem processes that affect vegetation communities modified by human activity.

Detailed chronologies were compiled for each site using a plotless sampling unit, i.e., a soil-landscape association, the location and general extent of which remains unchanged over time. The dimensions of the site are georeferenced as a centroid, which remains constant back in time, now and into the future.

For each site, we used the 10 criteria and 22 indicators (Table 1) of VAST-2 as a checklist to search for and compile relevant spatio-temporal sources of data and information over time to generate a systemic and comprehensive site history. Sources of information included: published and unpublished accounts, scientific surveys, long-term ecological monitoring sites, land manager interviews, remote sensing and public-private data archives [9]. Our literature review included what is known about the unmodified or reference state plant community type for each site, which is described by the same 10 criteria and 22 indicators. Indicators from the reference state were used in a relative sense to assess the transformation of each site over time.

We also used the 10 criteria and 22 indicators to assess the response of each plant community to the effects of the management practices. This process involved integrating and evaluating the site-based environmental histories and the response of the plant community over space and time. The integration of the relative difference between the transformation of a site and its reference state determined the relative effects that land management practices have had on vegetation condition and resilience over time. An aggregate index for each year in the chronology of a site is scored across four levels in a hierarchy (Table 1) [9].

We compiled and assessed the response of the plant community at each site in terms of structure, composition and function in response to land management regimes and practices. Therefore, it was necessary to define and describe land management regimes (or actions/interventions), as shown in Table 2. We then classified the responses of plant communities to these regimes based on how the practices of each regime individually and collectively transform the indicators of vegetation structure, composition and function over time. Collectively, the outcomes of these regimes are variously the maintenance, enhancement, restoration, degradation and or removal and replacement of a particular plant community at a site and or landscape.

We make a distinction between the reference state and a contemporary baseline. Most environmental monitoring and tracking of the responses of plant communities seek to measure and observe change relative to a current baseline. The VAST framework readily compiles and synthesises such data and information, where the attribute data being measured can be directly related to the fully-natural reference state for the criteria and indicators listed in Table 1.

Table 1. List of Vegetation Assets States and Transition (VAST) indicators, criteria and components of vegetation condition [9] used in this paper. Change is assessed relative to an assumed pre-European benchmark. A fourth level results in a vegetation status or transformation index derived by adding the weighted scores from Level 3.

Condition Components [1]	Key Functional, Structural and Composition Criteria	Indicators
Level 3	Level 2	Level 1
Functional	Soil hydrology	Rainfall infiltration and soil water holding capacity
		Surface and subsurface flows
	Soil physical status	Effective rooting depth of the soil profile
		Bulk density of the soil through changes to soil structure or soil removal
	Soil nutrient status	Nutrient stress: rundown (deficiency) relative to reference soil fertility
		Nutrient stress: excess (toxicity) relative to reference soil fertility
	Soil biological status	Organisms responsible for maintaining soil porosity and nutrient recycling
		Surface organic matter, soil crusts
	Natural disturbance regime	Area/size of disturbance events: foot prints (e.g., major storm cells, floods, wildfire, cyclones, droughts, ice)
		Interval between disturbance events
	Reproductive potential	Reproductive potential of overstorey structuring species
		Reproductive potential of understorey structuring species
Structural	Overstorey structure	Overstorey top height (mean) of the plant community
		Overstorey foliage projective cover (mean) of the plant community
		Overstorey structural diversity (i.e., a diversity of age classes) of the stand
	Understorey structure	Understorey top height (mean) of the plant community
		Understorey ground cover (mean) of the plant community
		Understorey structural diversity (i.e., a diversity of age classes) of the plant
Compositional	Overstorey composition	Densities of overstorey species functional groups
		Richness: the number of indigenous overstorey species relative to the number of exotic species
	Understorey composition	Densities of understorey species functional groups
		Richness: the number of indigenous understorey species relative to the number of exotic species

[1] Modified from the functional, structural and compositional levels of organization observed in biological diversity [17].

Table 2. Five land management regimes used to evaluate the response of a native plant community to land management practices, relative to the reference state.

Management Regimes
1. No active intervention that affects indicators of vegetation function, structure and composition
2. Management practices that harvest vegetation products (biomass, fibre, flowers, fruit and nuts), which affect indicators of vegetation function, structure and composition
3. Management practices that enhance or improve indicators of vegetation function, structure and composition
4. Management practices that extirpate or remove indicators of the function, structure and composition
5. Management practices that reconstruct or reinstate indicators of the function, structure and composition

Modified from [20].

The VAST system also presents a simple graphical report card showing the drivers of change and trend relative to a reference state (i.e., natural benchmark). Existing reference states were obtained from published sources or were elicited from skilled local ecologists and botanists [9]. The graph represents a transformation trajectory for a plant community where the condition (i.e., vegetation status) is scored out of a potential 100% (i.e., an unmodified reference state). The total score is comprised of three weighted components: function (regenerative capacity; 55% weighting); vegetation structure (27% weighting); and species composition 18% [9]. This weighting was applied in the same manner across all case studies. The total vegetation status score was calibrated to the six VAST classes [2], enabling the broad description of types of changes in condition over time. The degree of divergence between the reference state and the vegetation scores over time for each case study represents the degree of modification. Scores are grouped according to the following intervals:

80%–100% of the reference state corresponds to a residual/unmodified state;

60%–80% corresponds to a modified state;

40%–60% corresponds to a transformed state;

20%–40% corresponds to VAST Class IV: replaced and adventive; as well as

0%–20% corresponds to VAST Class V: replaced and managed; and VAST Class VI: replaced.

These five intervals provide a meaningful basis for describing and summarising change.

3. Results

The 250-year dynamics and drivers of the vegetation condition of case studies from the 10 agro-climatic regions of Australia are summarised in Table 3. These dynamics and drivers are graphically shown in Figures 2–11.

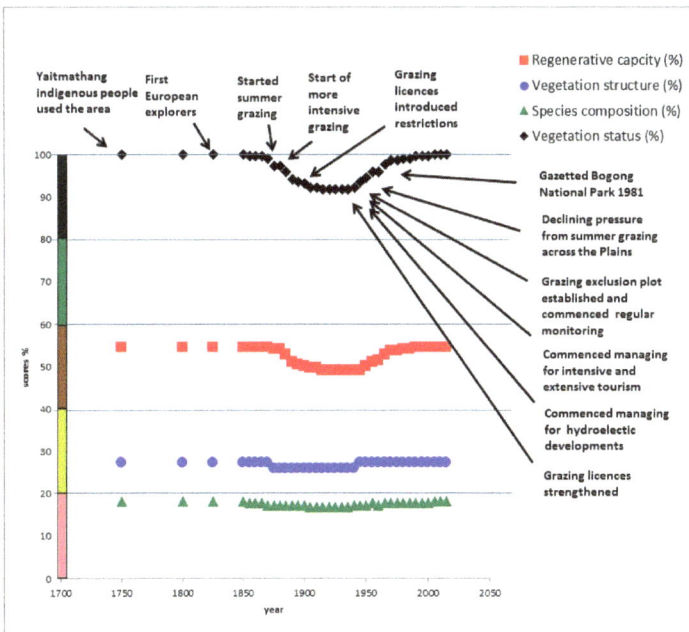

Figure 2. Case Study 1: Rocky Valley, Bogong High Plains, cold wet agro-climate region, Poa tussock grassland.

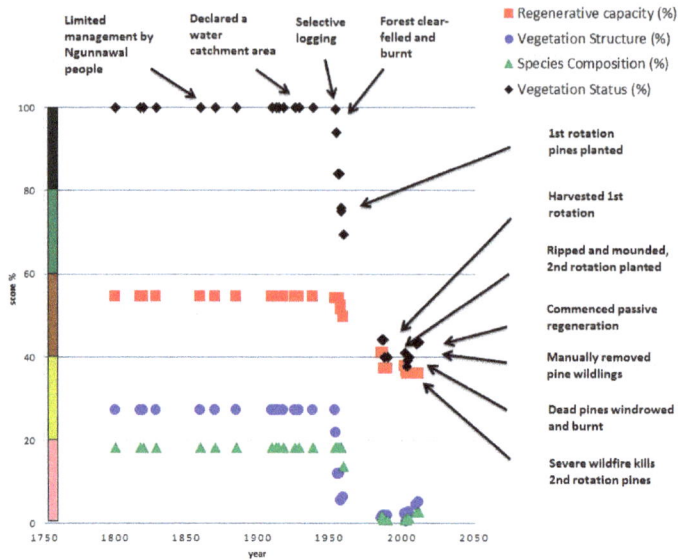

Figure 3. Case study 2: Blundells Flat, Brindabella Range, temperate cool-season wet agro-climate region, eucalypt open forest.

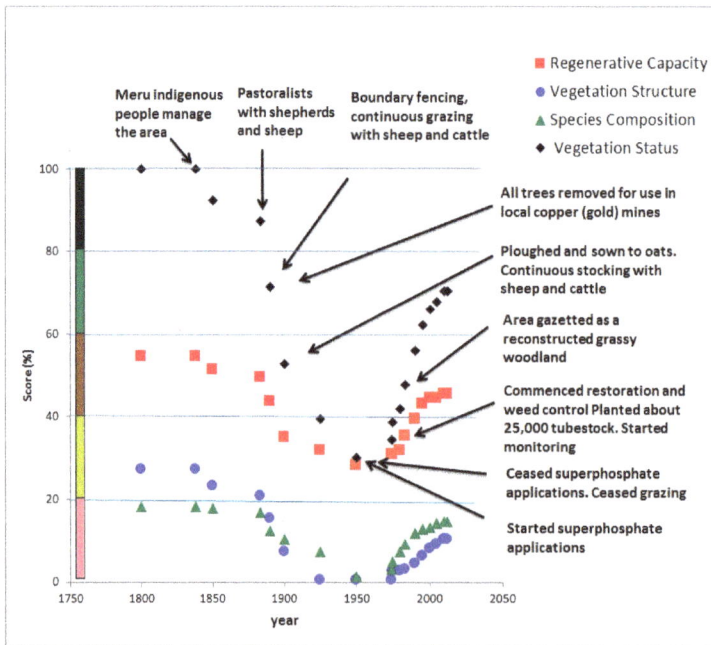

Figure 4. Case study 3: Wirilda, Harrogate, Mediterranean agro-climate region, *Callitris*, eucalypt low mallee woodland.

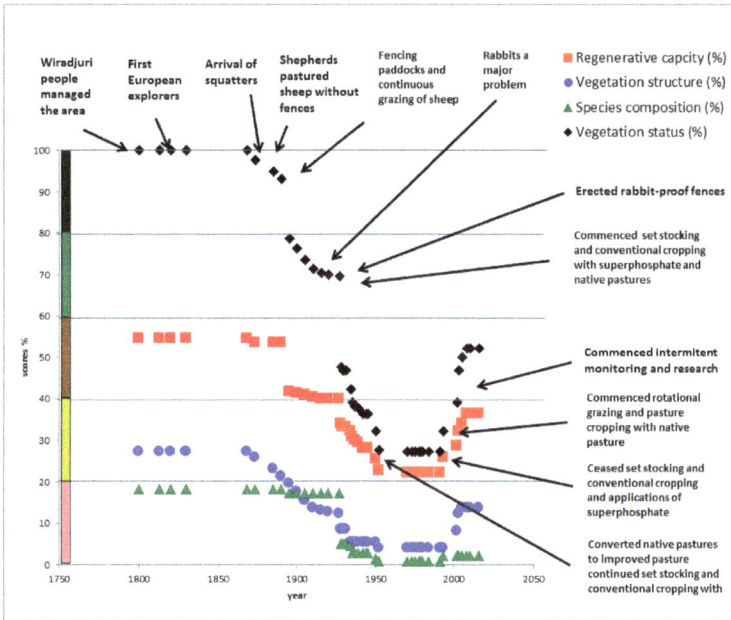

Figure 5. Case study 4: Winona, Gulgong, temperate, sub-humid agro-climate region, tussock grassland, eucalypt open woodland.

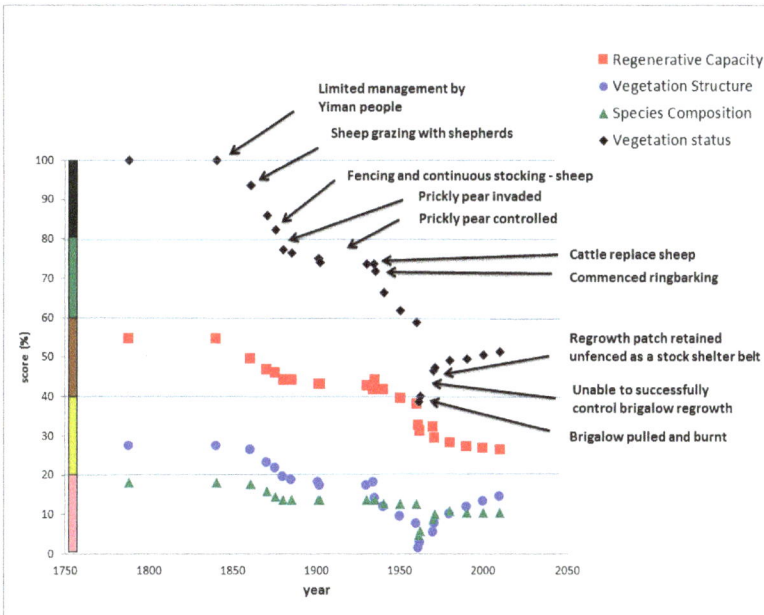

Figure 6. Case study 5: Potters Flat, Wandoan, sub-tropical sub-humid agro-climate region, *Acacia* open forest and woodland.

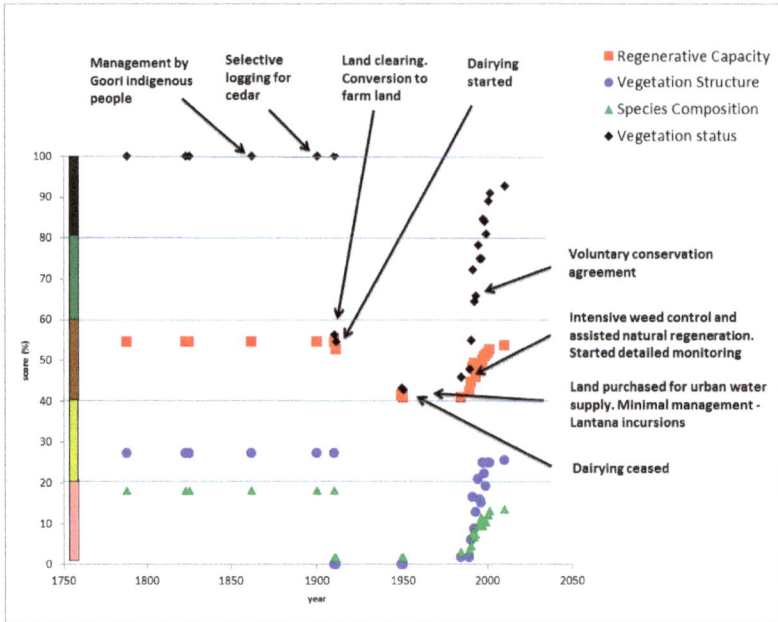

Figure 7. Case study 6: Rocky Creek Dam, Big Scrub, sub-tropical moist agro-climate region, lowland sub-tropical rainforest.

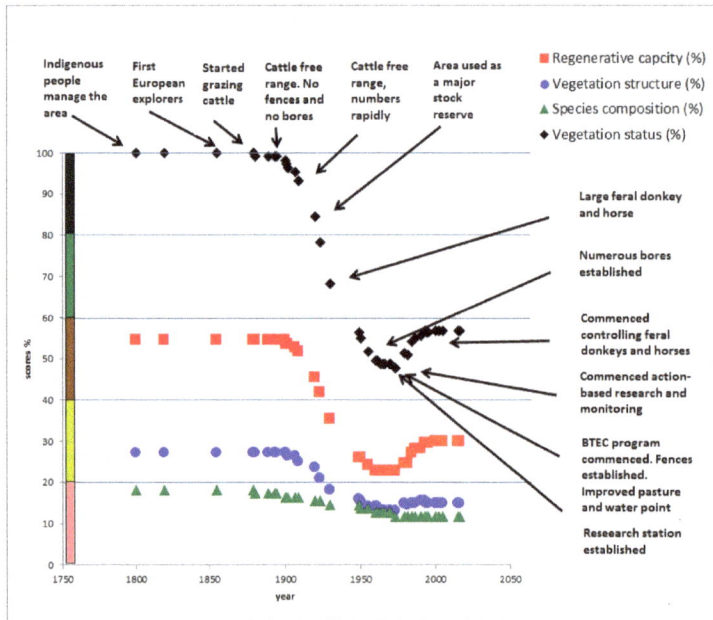

Figure 8. Case study 7: Conkerberry Paddock, Victoria River Research Station, tropical warm season wet agro-climate region, eucalypt open woodland.

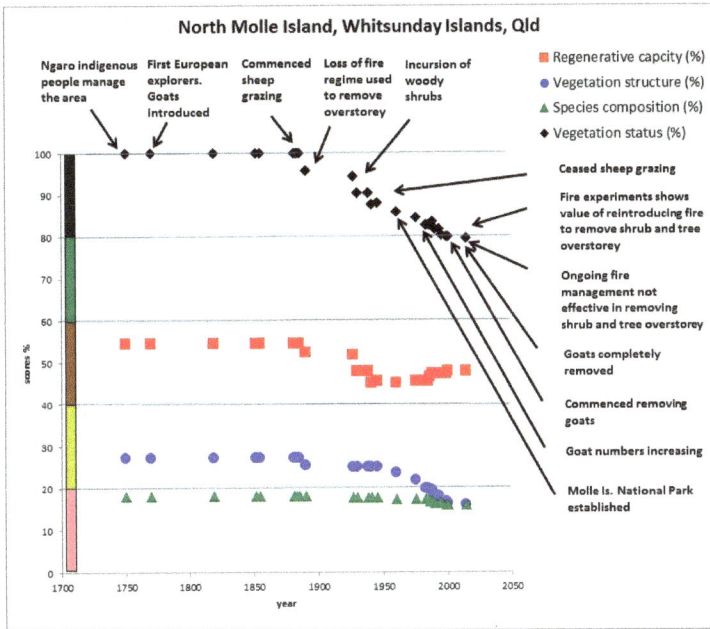

Figure 9. Case study 8: North Molle Island, Molle Group, Cumberland Islands, tropical warm season moist agro-climate region, Imperata and Themeda tussock grassland.

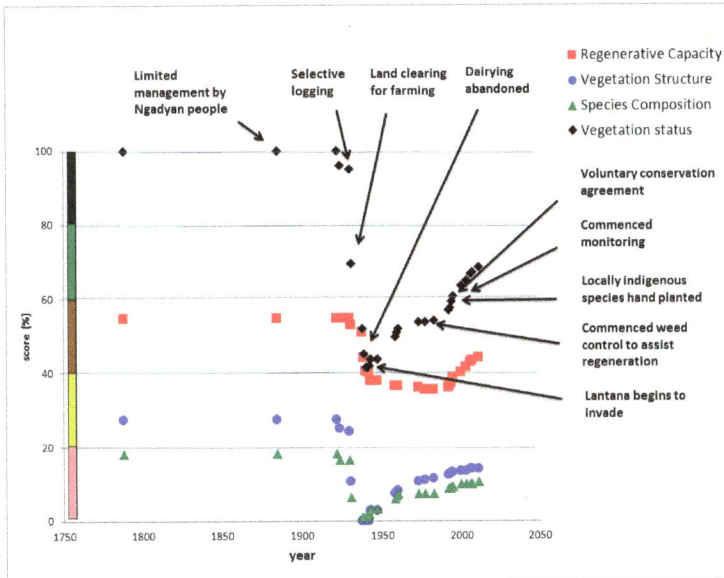

Figure 10. Case study 9: Wooroonooran Nature Refuge, tropical wet agro-climate region, complex mesophyll vine forest.

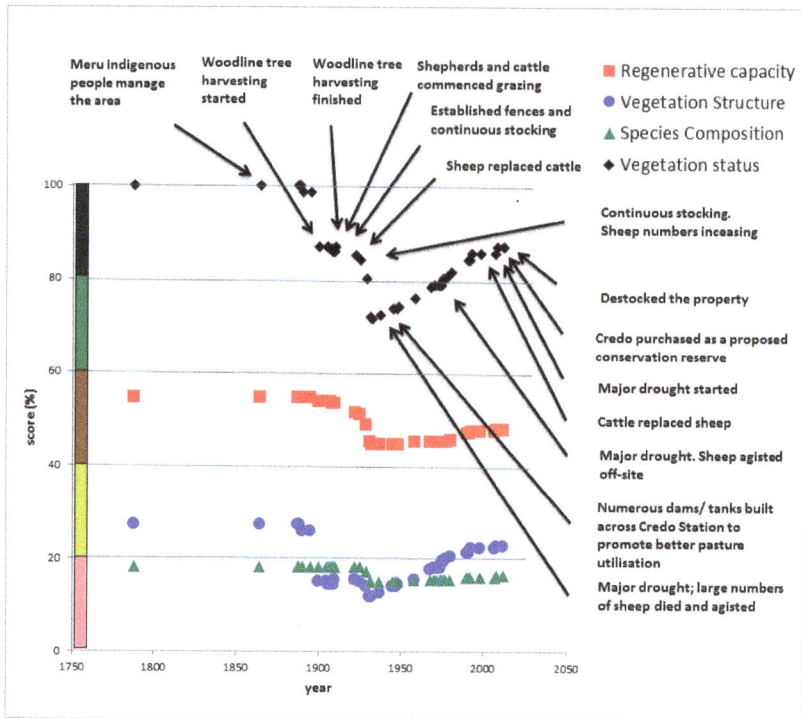

Figure 11. Case study 10: Chadwin paddock, Credo Station, dry agro-climate region, eucalypt woodland.

Detailed descriptions and explanations of each case study are provided in the Supplementary Material.

These results illustrate the dominant transformation pathways that have affected much of the Australian continent. These transformations include processes of replacement, removal and recovery of natural ecosystems; noting that native vegetation is used as an integrating surrogate for ecosystems. The responses of the 10 case studies illustrate different influences that individuals, local communities, government policies, markets and climate variation have had in reducing vegetation composition, structure and function. Though in half of the cases (Figures 2, 4, 5, 7 and 9), there has been recent and substantial restoration of composition, structure and regenerative capacity of the local native vegetation. The one common and profound driver of loss of vegetation condition for all of these case studies has been the rapid and near complete displacement of indigenous communities and land management practices by people, technology and land management practices of recent European origins.

Figures 2–11 show the total vegetation status score calibrated to VAST classes: 80%–100% of the reference state corresponds to VAST Class I: residual/unmodified (dark green bars); 60%–80% corresponds to VAST Class II: modified (mid-green bars); 40%–60% corresponds to VAST Class III: transformed (brown bars); 20%–40% corresponds to VAST Class IV: replaced and adventive (lime green bars); 1% to 20% corresponds to VAST Class V: replaced and managed (pink bars); and 0% corresponds to VAST Class VI: replaced and removed (red bars, not relevant in these case studies).

Table 3. Drivers of changes in vegetation conditions across representative case studies from Australia's ago–climatic zones based on Figures 2–11 and case study descriptions provided in Tables S1 and S2 in the Supplementary Material.

Case Study	Agro-Climatic Zone and Reference Vegetation	Vegetation Condition Dynamics (Status Score Change Relative to Reference State)	Government Policies	Markets	Technological Changes	Climate Variation	Cultural
1	Cold-wet, alpine grassland	10% loss due to livestock grazing, then recovery to near reference conditions	State government reduction, then prohibition of livestock grazing, then creation of a national park	Rapid development of national and international markets for meat and wool	Domestic livestock	Periods of drought that increased livestock grazing pressure on alpine grassland	Total indigenous displacement by Western European values and land management practices
2	Cold-wet, open forest	60% loss due to forest clearing, minor recovery by passive restoration	National and territory government-funded establishment of softwood plantations, then initiation of restoration of native vegetation for water catchment values	Domestic market for softwood for housing construction now influenced by softwood supply from New Zealand	*Pinus radiata* plantation system	Severe wildfire destroys pine plantation in 2003, linked to prolonged drought, as well as periods of above average rainfall that supported passive restoration (El-Nino–La Nina cycles)	Total indigenous displacement by Western European values and land management practices
3	Mediterranean, low (mallee) woodland	70% loss due to clearing, then significant recovery by active restoration	Various regulations that required clearing, then subsidies on fertiliser to increase intensification, more recently, agri-environment schemes to support farmers to restore native vegetation	Demand for timber for smelting of copper, then domestic and international demand for meat wool and grains	Mining technology, broad-scale cropping and exotic pasture systems, domestic livestock, fencing and feral rabbits	Periods of drought that hastened loss of vegetation condition, as well as periods of above average rainfall that supported restoration (El-Nino–La Nina cycles)	Total indigenous displacement by Western European values and land management practices
4	Temperate, sub-humid, grassy eucalypt woodland	70% loss due to clearing then partial recovery	Various regulations that required clearing, then subsidies on fertiliser to increase intensification, more recently, later agri-environment schemes to restore native vegetation	Domestic and international demand for grains, meat and wool	Broad-scale cropping and exotic pasture systems, domestic livestock, fencing and feral rabbits; no till cropping into dormant native pasture with cell-based sheep grazing	Drought and wildfire were a stimulus for land management change coupled with localised rising ground water that was saline.	Total indigenous displacement by Western European values and land management practices
5	Sub-tropical subhumid, *Acacia* forest and woodland	60% loss due to clearing and introduction of exotic pasture grasses, weeds, then small-scale recovery by *Acacia* regrowth	Various regulations that required clearing by land owners	Domestic and international demand for grains, meat and wool	Broad-scale mechanical clearing, cropping and exotic pasture systems, domestic livestock and fencing	Unknown impact of climate variation	Total indigenous displacement by Western European values and land management practices
6	Sub-tropical moist, lowland rainforest	60% loss due to land clearing and conversion to exotic pastures, colonisation by a woody weed, then significant recovery due to active restoration	Various regulations that required clearing by land owners, then initiation of restoration of native vegetation for water catchment and local eco-tourism values	Domestic and international demand for sub-tropical timbers, then domestic demand for dairy products	Fencing for intensive dairy production, then development of the science and practice of ecological restoration	An area of less climatic variation than inland Australia	Total indigenous displacement by Western European values and land management practices

Table 3. *Cont.*

Case Study	Agro-Climatic Zone and Reference Vegetation	Vegetation Condition Dynamics (Status Score Change Relative to Reference State)	Government Policies	Markets	Technological Changes	Climate Variation	Cultural
7	Tropical warm season wet, eucalypt open woodland	50% loss due to livestock and feral herbivore grazing with modest recovery due to improved grazing management and increasing woody cover due to climate change	Government-managed livestock reserve, and subsidies for artificial watering points and fencing, then R&D into improved range management	Domestic and particularly international demand, including live cattle exports	Artificial watering points (bores), then fencing to improve grazing management, improved roads and transport, introduction of *Bos indicus* breeds of cattle	Large seasonal fluctuations in rainfall affecting livestock and feral herbivore numbers, but overall increasing rainfall over a longer season	Total indigenous displacement by Western European values and land management practices; a conditional land claim was granted in 1990, enabling continued use of the area as a research station
8	Tropical warm season moist, tussock grassland	20% loss due to loss of Indigenous fire regime that controlled woody cover	National park status declared in 1938, eventual removal of domestic and feral goat grazing pressure	Limited use by domestic livestock for local consumption	Aerial incendiaries applied, but with limited success in reducing woody cover	Limited impact of seasonal variations	Total indigenous displacement by Western European values and land management practices
9	Tropical wet, vine forest	60% loss due to land clearing, then moderate recovery by passive and active restoration	State government land development policies that promoted land clearing for dairy, but more recently, site declared a nature refuge, providing public and private benefits	Initially demand for high value tropical timbers, then domestic dairy production and subsequent collapse due to high costs	Introduction of exotic pasture grasses and dairy production system, including fertilizers and lime, then modest demand for 'life-style' blocks of land with new owners passionate about restoration	Very high annual rainfall accelerated soil erosion and fertility decline, though this rainfall also supported rapid ecological restoration	Total indigenous displacement by Western European values and land management practices
10	Dry, eucalypt woodland	30% loss due to timber harvesting and livestock grazing, then modest recovery due to improve grazing management	Recently purchased by state government as a proposed conservation reserve	Global demand for minerals requiring timber for smelting, then demand for meat and wool	Smelting technologies requiring timber for fuel, fencing, artificial watering points, domestic livestock	Periods of drought requiring artificial sources of water for livestock (bores and troughs)	Total indigenous displacement by Western European values and land management practices

4. Discussion

We ambitiously set ourselves the task of describing and understanding circa 250 years of landscape transformation of native ecosystems from case studies in 10 contrasting ago-climatic regions of Australia. We have done so by using VAST, a robust framework and methodology that broadly quantifies changes in native vegetation composition, structure and function. These regionally-distinct case studies demonstrate that the condition of vegetation at key points in time is an emergent property of economic markets, new technologies, the history of settlement, environmental constraints, government policies and programs and the impact of individual land manager's practices and values. These studies illustrate the interaction of people living in, learning from and imperfectly adapting to their environment. We suggest that this systematic application of the VAST framework provides rigorous and consistent information for land use planners, policy makers and land managers to design and apply appropriate interventions to improve the vegetation condition and delivery of a diversity of ecosystem services.

We further discuss the significance of these representative case studies in regards to the importance of multiple scales of interpretation, drivers of change and the importance of recognising the resilience status of a given landscape at a given time. These issues relate to the landscape management principles developed by Sayer and others [5].

4.1. Emergent Impacts at Multiple Scales

Understanding how a site is transformed spatially over time provides critical insights for a diversity of stakeholders to address the full spectrum of human impacts observed across modified and fragmented landscapes. The vegetation condition framework we have applied is a site-based concept [2], whereas landscape alteration levels are an emergent property of these finer scale representations of vegetation condition [3]. Levels of landscape alteration represent the aggregate of varying degrees of landscape fragmentation and increasing degrees of site modification [1–3]. A framework for conceptualizing the effects of landscape fragmentation and increasing degrees of site modification and understanding their relevance to management, as shown in Figure 12, has been widely accepted [1]. Each of the 10 case studies described above are set within this framework to illustrate how the 10 case studies are set on a pathway toward increasing modification and fragmentation or have been reset towards a pathway of decreasing modification and fragmentation.

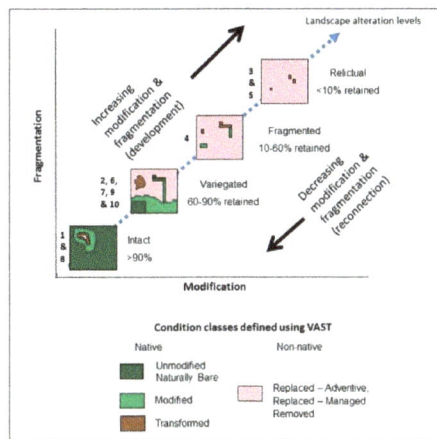

Modified from [21]

Figure 12. Case Studies 1–10 shown in the context of a gradient of landscape alteration levels [3] and VAST classes [2].

In the early stages of rural development all case studies occurred in an intact landscape context (within the dark green bars of Figures 2–11). Case Studies 1 and 8 (Figures 2 and 9) represent intact landscapes where greater than 90% of the mapped extents of vegetation condition are retained comprising three condition classes, unmodified, modified and transformed [2]. The bulk of intact landscapes (Figure 12) are represented by unmodified condition classes. Case Studies 2, 6, 7, 9 and 10 are found in variegated landscapes where 60%–90% of the native vegetation is retained comprising three condition classes, unmodified, modified and transformed [2]. The bulk of the variegated landscapes is comprised of the modified condition class (mid-green bars shown in Figures 2–11). Case Study 4 is found in fragmented landscapes, retaining only 10%–60% of the native vegetation [2]. Case Studies 3 and 5 are found in relictual landscapes, where less than 10% of the native vegetation is retained [2].

Depicting the 10 case studies using such a landscape model (Figure 12) has a benefit because it provides a policy and planning context for developing and implementing public-private natural resource management programs and partnerships. Particular classes of landscape fragmentation and modification, representing emergent properties of finer scale vegetation condition states, can be identified and prioritised. Such prioritisation can target land managers and local communities with incentives to change land management practices to enhance the indicators of vegetation condition.

This conceptual landscape model can also be used to guide the monitoring and reporting of change at the site and landscape levels to evaluate and track the outcomes of public-private incentive programs. Australian examples of public-private programs that are actively engaging private landholders to relink fragmented landscapes and improve the extent and condition of modified and degraded ecosystems have been described [22]. While such public-private programs vary between regions, we have demonstrated that the case study sites retain varying degrees of resilience in terms of vegetation structure, composition and function. Our results show that many of the broader natural resource management issues pertaining to most of the Australian agro-climatic regions [13] are being confronted and partially resolved least at the site scale. Many land managers working within these regions are beginning to demonstrate a knowledge of ecosystem dynamics to understand, value and restore vegetation condition and ecosystem resilience [22].

4.2. Major Drivers of Change

In all 10 case studies, the main drivers of environmental change are a complex interplay of social, economic and environmental factors impacting how sites and landscapes have been managed and transformed (Table 3). Throughout the history of landscape development and the recovery of landscapes in Australia, state and national governments and, more recently, regional natural resource management bodies have played a major role in influencing how sites and landscapes are managed and transformed. Local communities are one player influencing how the land is managed, albeit a relatively minor player. The exception to this is pattern is Landcare, a community-based movement, which is involved in identifying and resolving natural resource management issues associated with managing land for agricultural productivity [23]. It is worth noting that in developing the chronologies of change in resilience of the 10 case studies, no records of the contribution of the Landcare movement were discovered.

In the early stages of agricultural development in Australia, first British colonial, then state government agencies were largely responsible for commanding or facilitating rural resource exploitation, particularly agricultural and forestry land uses, into areas that were previously all occupied and managed by a diversity of indigenous nations. This was the case regarding the early allocation of land for grazing (e.g., Figures 2 and 4, Figures 5–11); the development of the Brigalow lands for intensive agriculture (e.g., Figure 6); and the allocation of land for forest plantations (e.g., Figure 3).

While there are instances where individuals and local communities moved ahead of the government's ability to control rights and access in developing new areas, as in the case of gold fossickers, miners and squatters, colonial, then state governments have generally initiated and controlled access to new areas for agricultural development. For examples where individuals and local communities occupied, managed and transformed landscapes, it can be argued that their impacts were either low and short in duration, as in the case of squatters, or high impact, but confined to small areas, as in the case of mining.

Arguably, over the longer term, market forces and technologies have had a much greater impact on how the land was managed and transformed. Revenue gained from domestic and export sales of agricultural, timber and mining products has been used to apply increasingly sophisticated technologies (e.g., from axes to bulldozers) to modify the vegetation structure, composition and function of sites and landscapes.

4.3. Resilience

Collectively, the impact of land management practices and technologies that are market driven or promoted by government policies (e.g., subsidies) has profoundly reduced the vegetation condition and the ecosystem resilience across much of Australia [4,21]. This is particularly the case where intensive agriculture and forestry have removed and replaced the native overstorey or replaced the understorey structure and composition with exotic trees, pastures and crops (e.g., Figures 3–5, 7 and 10). Intensive agriculture and forestry have modified key functional criteria of soil hydrology, soil nutrients, soil structure, soil biology, the natural disturbance regime and the reproductive potential of the plant community (Table 1). These ecosystem changes have, at least in the short term (decades), promoted higher levels of economic productivity than would otherwise be the case under a cover of native vegetation. For example, applications of superphosphate (e.g., Figures 4 and 5) have been used to modify the soil nutrients to promote plant productivity; mechanical cultivation has been used to reduce soil bulk density (soil compaction) and improve rainfall infiltration and water holding capacity. Clearing, prolonged cultivation and the application of herbicides have been used to limit the regrowth of native vegetation to promote agricultural crops and pastures. However, this prolonged disturbance has reduced the natural regenerative capacity of native vegetation to the point that active and costly restoration of native vegetation is now required (e.g., Figures 4 and 10).

In some case studies, market forces, government policies or individual landholder values have changed such that agricultural intensification has been reduced. This has had profound and positive impacts on stimulating natural (spontaneous) regeneration and improving vegetation condition, but only in especially those ecosystems that have retained high levels of resilience. This is the case with the cessation of particular rural industries, e.g., dairying (Figures 7 and 10), pine plantations (e.g., Figure 3) or livestock grazing (e.g., Figures 2 and 9). Cessation of these industries has at least modestly increased the structure, composition and function towards reference states of the native vegetation. The exception was Case Study 8 (Figure 9), where the cessation of domestic and feral livestock grazing has not resulted in increases in vegetation condition towards the pre-European reference state because an alternate reference state (woody cover) has been adopted as a goal of government conservation management.

To a lesser extent, where the land use has been maintained, but the intensity of land management practices has been reduced, there has been a significant increase in the indicators of vegetation condition. For example, within Case Study 4 (Figure 5), a conventional farming system has been replaced with rotational grazing and low intensity cropping into native pasture, which has significantly increased vegetation structure and regenerative capacity, but has not increased the species composition of this formally wooded landscape. In Case Study 7 (Figure 8), grazing pressure by cattle and feral donkeys has been reduced, but not eliminated, resulting in a significant gain in the regenerative capacity of the natural grass pasture, but little influence on vegetation structure and composition. Similarly, for Case Study 10 (Figure 11), fencing and numerous artificial watering points have resulted in less intensive sheep and cattle grazing, stimulating a modest recovery of regenerative capacity, but little impact on vegetation composition and structure.

All of the case studies exhibited dynamic responses reflecting the interaction between natural and anthropogenic processes; hence, each resilience assessment is time bound [4]. The above case studies can be ordered from highest to lowest, based on the current resilience and management regimes, that is the relative difference between the reference state and recovery of the transformation index. Case studies shown in Figures 2, 7, 9 and 11 are sites that have the highest resilience (80–100%), Figures 4 and 10 with mid-range resilience (60–80%) and Figures 3, 5, 6 and 8 with the lowest resilience (40–60%). This kind of analysis helps to diagnose and understand the effects of land management practices on the response of the plant community as represented by the trajectories of the 22 indicators (Table 1) over time.

This analysis of resilience helps to develop prognoses based on the current trend of the condition index scores for the site relative to timing and likely the effectiveness of management interventions; for example, identifying sites that have moderate to high resilience (e.g., high regenerative capacity, Figure 7) that are amenable to passive strategies, such as reduced grazing and weed control to allow natural tree regeneration, which is usually far less expensive than active tree planting. In contrast, active and expensive restoration practices should be targeted to sites that have been recognised to have lost their regenerative capacity, e.g., Figure 4.

4.4. Changing Values and Attitudes

Changing social values and attitudes of individuals, communities and governments over the past 25 years or so have resulted in the improvement in vegetation condition within most of the examined case studies. We have documented diverse relationships between public land use policy and planning, land manager's practices and the responses of plant communities over time. These results indicate that in many cases, there are complex relationships between government policy, local communities and private land managers in regards to how sites are managed. For example, the Queensland Government's decision in the 1980s not to reinstate a fire management regime appropriate for restoring the reference state on North Molle Island (Imperata and Themeda tussock grassland) (Figure 9) was strongly influenced by park visitor surveys regarding community concerns over the burning vegetation on the islands in the Great Barrier Reef National Park [24–28]. Likewise, the Victorian Government decisions to control and remove grazing of cattle and sheep from the Bogong High Plains, Victoria (Figure 2), in the 1960s was strongly influenced by an increasing community awareness and concern of the impacts of grazing on alpine vegetation. This government decision was also influenced by increasing recreational demands for access to natural areas, increasing concerns about loss of biodiversity and adverse impacts on soil erosion and water harvesting [29,30]. Equally, the Australian Capital Territory decision not to re-establish a pine plantation at Blundell's flat (Figure 3), in the mid-2000s, was strongly influenced by community concerns over the likelihood of further wildfires; its severity appeared to be influenced by the flammability of pine plantations. In addition, pine plantations were not re-established in a major water catchment because of community concerns over the effect of plantations on water quality and quantity [31].

These three examples illustrate the roles community and government have in influencing changes in vegetation structure, composition and function. While the spatial extent of these case studies is small, they represent more broadly changes in community attitudes and values that can be observed well beyond these case studies. Changing community attitudes and values have seen an increased demand for the creation and conservation of natural areas resulting in the removal of grazing from almost all formally protected public and private nature reserves across Australia. In the early colonial development of Australia, the prevailing community expectation was of privately managed rural enterprises supported by government policies and incentives to support food and timber industries and the national economy. Since about the late 1930s, community attitudes and values have moved away from solely supporting agricultural and forestry enterprises, towards conserving natural areas for public benefits and services.

5. Conclusions

We hope to have demonstrated that representative and systematic site-level assessments of vegetation condition dynamics over hundreds of years can provide valuable insights into how landscapes have been altered. This long-term perspective can then be used to help identify interventions to meet current and future community values and expectations. We have tried to demonstrate that this kind of rapid, but repeatable analysis can help answer such questions as: What is the condition of the native vegetation at a site relative to an accepted national standard? How can one assess and report consistently and at multiple scales the condition of ecosystems resulting from management interventions? As a land manager, how can this knowledge be used to improve the condition of a site or landscape? These are questions that need a broad, consistent and historical approach to answering.

We suggest that the kind of visual presentation of the results shown in this paper allows policy makers and land managers to quickly recognise and understand how complex socio-ecological processes can affect ecosystem dynamics and services. Such graphically-presented information should also be useful for monitoring, analysing and reporting the effects of land management practices at the selected sites [32,33], as well as for regional accounts of native vegetation condition [34].

We have tried to demonstrate that a systemic approach to documenting and explaining the historical dynamics of vegetation composition, structure and regenerative capacity is useful for tracking and evaluating the diverse drivers of change. Understanding these drivers is critical to developing an ecological and social understanding of how vegetation condition and ecosystem services at multiple scales can be best managed into the future.

Supplementary Materials: The following are available online at www.mdpi.com/2073-445X/5/4/40/s1, Table S1 List of 10 sites assessed using the VAST-2 criteria and indicators of structure, species composition and function (Table 1) showing the associated agro-climate regions, bioregions and plant communities, Table S2. Descriptions of agro-climatic regions, Case Studies 1–10.

Acknowledgments: We thank each of the land managers and researchers who provided detailed records on the history of the management of each site. Their contribution has enabled the effects of land management and the responses of the respective plant communities to become part of the public record, thus demonstrating the value of compiling and integrating qualitative spatial and temporal information with quantitative sources of information.

Author Contributions: Richard Thackway and David Freudenberger conceived and designed the experiments; performed the experiments; analysed the data; contributed reagents/materials/analysis tools; and wrote the paper.

Conflicts of Interest: The authors declare no conflict of interest. The founding sponsors had no role in the design of the study; in the collection, analyses, or interpretation of data; in the writing of the manuscript, and in the decision to publish the results.

References

1. McIntyre, S.; Hobbs, R.J. Human impacts on landscapes: Matrix condition and management priorities. In *Nature Conservation 5: Nature Conservation in Production Environments: Managing the Matrix*; Craig, J.L., Mitchell, N., Saunders, D.A., Eds.; Surrey Beatty and Sons: Chipping Norton, Australia, 2000; pp. 301–307.
2. Thackway, R.; Lesslie, R. Describing and mapping human-induced vegetation change in the Australian landscape. *Environ. Manag.* **2008**, *42*, 572–590. [CrossRef] [PubMed]
3. Mutendeudzi, M.; Thackway, R. *A Method for Deriving Maps of Landscape Alteration Levels from Vegetation Condition State Datasets*; Bureau of Rural Sciences: Canberra, Australia, 2010.
4. Resilience Alliance. *Assessing Resilience in Social-Ecological Systems: Workbook for Practitioners*, Version 2.0; 2010. Available online: http://www.resalliance.org/3871.php (accessed on 31 January 2016).
5. Sayer, J.; Sunderland, T.; Ghazoul, J.; Pfund, J.L.; Sheil, D.; Meijaard, E.; Venter, M.; Boedhihartono, A.K.; Day, M.; Garcia, C.; et al. Ten principles for a landscape approach to reconciling agriculture, conservation, and other competing land uses. *Proc. Natl. Acad. Sci. USA* **2013**, *110*, 8349–8356. [CrossRef] [PubMed]
6. Woollacott, A.; Adcock, M.; Allen, M.; Evans, R.; Mackinnon, A. *Overview: The Making of the Modern World (1750–1918)*; Cambridge University Press: Cambridge, UK, 2010.
7. Mulvaney, D.J. *The Prehistory of Australia*; Frederick A Praeger: New York, NY, USA, 1969.

8.   Blainey, G. *The Story of Australia's People. The Rise and Fall of Ancient Australia*; Viking: Melbourne, Australia, 2015.

9.   Thackway, R.; Specht, A. Synthesising the effects of land use on natural and managed landscapes. *Sci. Total Environ.* **2015**, *526*, 136–152. [CrossRef] [PubMed]

10. Brown, V.A.; Lambert, J.A. *Collective Learning for Transformational Change: A Guide to Collaborative Action*; Routledge: London, UK; New York, NY, USA, 2013.

11. Yapp, G.; Walker, J.; Thackway, R. Linking vegetation type and condition to ecosystem goods and services. *Ecol. Complex.* **2010**, *7*, 292–301. [CrossRef]

12. Schaffer, G.; Levin, N. Mapping human induced landscape changes in israel between the end of the 19th century and the beginning of the 21th century. *J. Landsc. Ecol.* **2014**, *7*, 110–145. [CrossRef]

13. Hobbs, R.J.; McIntyre, S. Categorizing Australian landscapes as an aid to assessing the generality of landscape management guidelines. *Glob. Ecol. Biogeogr.* **2005**, *14*, 1–15. [CrossRef]

14. Hutchinson, M.F.; McIntyre, S.; Hobbs, R.J.; Stein, J.L.; Garnett, S.; Kinloch, J. Integrating a global agro-climatic classification with bioregional boundaries in Australia. *Glob. Ecol. Biogeogr.* **2005**, *14*, 197–212. [CrossRef]

15. Williams, J.; Hook, R.A.; Hamblin, A. *Agro—Ecological Regions of Australia, Methodologies for Their Derivation and Key Issues in Resource Management*; CSIRO Land and Water: Canberra, Australia, 2002.

16. Köppen, W. *Die Klimate der Erde*; De Gruyter: Berlin, Germany, 1923.

17. Noss, R.F. Indicators for Monitoring Biodiversity: A Hierarchical Approach. *Conserv. Biol.* **1990**, *4*, 355–364. [CrossRef]

18. Tropical Savannas Cooperative Research Centre. *Savanna Explorer Northern Australian Information Service.* Retrieved from The Population of Savanna Australia: (n.d.). Available online: http://www.savanna.org.au/all/economic.html (accessed on17 April 2016).

19. Fensham, R.J.; Fairfax, R.J.; Archer, S.R. Rainfall, land use and woody vegetation cover change in semi-arid Australian savanna. *J. Ecol.* **2005**, *93*, 596–606. [CrossRef]

20. Thackway, R. *Tracking Anthropogenic Influences on the Condition of Plant Communities at Sites and Landscape Scales*; Almusaed, A.Z., Ed.; Landscape Ecology: Rijeka, Croatia, 2016.

21. Yapp, G.A.; Thackway, R. Responding to change—Criteria and indicators for managing the transformation of vegetated landscapes to maintain or restore ecosystem diversity. In *Biodiversity in Ecosystems—Linking Structure and Function*; Blanco, J.A., Ed.; InTech: Rijeka, Croatia, 2015.

22. Fitzsimons, J.; Pulsford, I.; Wescott, G. (Eds.) *Linking Australia's Landscapes, Lessons and Opportunities from Large-Scale Conservation Networks*; CSIRO Publishing: Collingwood, Australia, 2013.

23. Brown, V.A. *Landcare Languages: Talking to Each Other about Living with the Land*; National Landcare Program, Department of Primary Industry: Canberra, Australia, 1995.

24. Queensland Herbarium. *Regional Ecosystem Description Database (REDD)*; Version 9.0; Queensland Department of Science, Information Technology and Innovation: Brisbane, QLD, Australia, 2015. Available online: https://environment.ehp.qld.gov.au/regional-ecosystems/ (accessed on 17 April 2016).

25. McDonald, K.; Atherton, Queensland. Personal communication, 2016.

26. Stanton, P.; Redlynch, Queensland. Personal communication, 2015.

27. Department of National Parks, Recreation, Sport and Racing. *Whitsunday Islands National Parks and Adjoining State Waters Management Statement 2013*; Department of National Parks, Recreation, Sport and Racing, Queensland Government, 2013. Available online: http://www.nprsr.qld.gov.au/managing/plans-strategies/statements/pdf/whitsunday-islands.pdf (accessed on 15 May 2016).

28. Stanton, P.; Stanton, D.; Stott, M.; Parsons, M. Fire exclusion and the changing landscape of Queensland's Wet Tropics Bioregion 1. The extent and pattern of transition. *Aust. For.* **2014**, *77*, 51–57. [CrossRef]

29. Victorian Conservation Trust. *Bogong High Plains, Vegetation Map and Guide to Alpine Flora. 1:15,000 Scale. One of a Series. Rocky Valley Sheet Including Falls Creek, Mt Nelse and Mt Cope Areas*; Victorian Conservation Trust in Conjuction with the Soil Conservation Service: Melbourne, Australia, 1986.

30. Australian Alps National Parks. *Grazing in the Australian Alps. Educational Resource.* Available online: https://theaustralianalps.files.wordpress.com/2013/11/grazing.pdf (accessed on 31 January 2016).

31. Thackway, R. *Blundells Flat, Ex-Coupe 424, ACT. Ver. 1. VAST-2: Tracking the Transformation of Australia's Vegetated Landscapes*; Australian Centre for Ecological Analysis and Synthesis, University of Queensland: Brisbane, Australia, 2012.

32. Montreal Implementation Group National Forest Inventory Steering Committee. *Australia's State of the Forests Report 2013*; Montreal Process Implementation Group for Australia and National Forest Inventory Steering Committee: Canberra, Australia, 2013.

33. Ecosystem Science Long-Term Plan Steering Committee. *Foundations for the Future: A Long-Term Plan for Australian Ecosystem Science*; Terrestrial Ecosystem Research Network, the Ecological Society of Australia, and the Australian Academy of Science's National Committee for Ecology, Evolution and Conservation: Canberra, Australia, 2014.

34. Sbrocchi, C.; Davis, R.; Grundy, M.; Harding, R.; Hillman, T.; Mount, R.; Possingham, H.; Saunders, D.; Smith, T.; Thackway, R.; et al. *Technical Analysis of the Australian Regional Environmental Accounts Trial*; Wentworth Group of Concerned Scientists: Sydney, Australia, 2015.

Article

Short-Term Projects versus Adaptive Governance: Conflicting Demands in the Management of Ecological Restoration

Ian Hodge [1],* and William M. Adams [2]

[1]   Department of Land Economy, University of Cambridge, 19 Silver Street, Cambridge CB3 9EP, UK
[2]   Department of Geography, University of Cambridge, Downing Place, Cambridge CB2 3EN, UK;
       wa12@cam.ac.uk
*   Correspondence: idh3@cam.ac.uk; Tel.: +44-1223-337-134

Academic Editors: Jeffrey Sayer and Chris Margules
Received: 2 August 2016; Accepted: 2 November 2016; Published: 10 November 2016

Abstract: Drawing on a survey of large-scale ecological restoration initiatives, we find that managers face contradictory demands. On the one hand, they have to raise funds from a variety of sources through competitive procedures for individual projects. These projects require the specification of deliverable outputs within a relatively short project period. On the other hand, ecologists argue that the complexity of ecosystem processes means that it is not possible to know how to deliver predetermined outcomes and that governance should be adaptive, long-term and implemented through networks of stakeholders. This debate parallels a debate in public administration between New Public Management and more recent proposals for a new approach, sometimes termed Public Value Management. Both of these approaches have strengths. Projectification provides control and accountability to funders. Adaptive governance recognises complexity and provides for long-term learning, building networks and adaptive responses. We suggest an institutional architecture that aims to capture the major benefits of each approach based on public support dedicated to ecological restoration and long-term funding programmes.

Keywords: ecological restoration; biodiversity conservation; adaptive governance; projectification; New Public Management; Public Value Management

1. Introduction

In the UK, large-scale conservation initiatives are being developed by a range of organisations in response to growing concern for landscape-scale ecological patterns and processes and interest in ecological restoration as a conservation strategy [1,2]. The shift of emphasis towards a larger scale has been promoted by the recognition in ecological thinking of the importance of the interconnectedness of areas of habitat at the landscape level. From this basis, the idea grew that conservation should be pursued through sets of protected areas, managed as part of 'ecological networks' (e.g., [3,4]). The approach was given strong support by the Lawton committee which concluded that existing nature reserves and designated wildlife sites in England did not form a 'coherent and resilient' ecological network [1]. Areas of highest conservation value were small and widely separated ('highly fragmented') and unsuited to coping with pressures such as climate change and economic growth. The report argued that "we need a step-change in our approach to wildlife conservation, from trying to hang on to what we have, to one of large-scale habitat restoration and recreation, under-pinned by the re-establishment of ecological processes and ecosystem services, for the benefits of both people and wildlife". This has led to a greater emphasis on projects aiming to implement ecological restoration at a larger scale than in the past [2].

This paper focuses on the contradictory demands on ecosystem restoration practitioners in light of a survey of restoration projects and the literature on New Public Management and projectification. We first draw on a survey of large-scale conservation initiatives in the UK [5,6] that highlights the pressures arising from short-term funding arrangements and the aspirations for a more adaptive approach towards management. The survey was undertaken in order to provide information on the ways in which such initiatives are being planned and managed. It identified a series of challenges in responding to the short-term requirements of the funding regimes, a process referred to as projectification [7], while at the same time seeking to maintain consistent long-term adaptive approaches to land management. Then, after considering the implications of the short-term project based funding, the paper examines the arguments for adaptive governance. The critique of projectification here parallels similar critiques of New Public Management in public administration. Sjöblom [8] has commented that "The gradual development towards increasingly non-permanent and informal structures is, in fact, one of the most important–although still very much neglected–administrative changes of the past decades." The paper then considers the relevance of moves proposed in that literature towards Public Value Management for conservation planning. Both approaches have aspects that have the potential to make a positive contribution to the effective implementation of ecological restoration initiatives. In this context, the paper seeks to integrate the different strengths of projectification and adaptive governance in the form of a thought experiment. This explores the potential for a new architecture for the implementation of ecological restoration, combining the opportunity for adaptive governance with the incentives and accountability provided through a project-based approach.

2. The Implementation of Ecological Restoration in the UK

Ecological restoration requires long-term control of land management to allow time for ecological processes and associated habitats to become established [9]. Larger-scale restoration generally requires the coordination of management beyond the borders of existing conservation areas, involving a range of landowning partners, including state and non-state actors (such as private landowners and managers and non-governmental conservation organisations and trusts) [2]. In consequence, fragmentation and institutional inefficiency constrain landscape-scale ecological management and restoration [10]. Active intervention into ecosystems for conservation purposes requires the investment of resources that have opportunity costs. Sponsors of restoration projects demand that funds are used efficiently and increasingly funds for conservation are allocated and managed through the mechanism of specific and relatively short-term projects [11]. A project may be defined simply as "a single intervention characterized by a fixed time schedule and dedicated budget" [8]. This implies control over the allocation and expenditure of funds by the organisation that controls the budgets. The literature suggests projects share a number of characteristics [12] that are evident in the implementation of ecological restoration:

- Involve a unique, once-in-a-lifetime task;
- Have a predetermined time frame;
- Are subject to one or several performance goals (such as resource usage and outputs);
- Involve a number of complex and/or interdependent activities.

Large-scale ecological restoration in the UK is typically undertaken by independent conservation organisations or consortia of organisations (often led by non-governmental conservation trusts), supported by one or more funders, which may be a charitable fund, a private firm or a government agency. The shift towards large-scale conservation initiatives means that conservation actions are increasingly undertaken in the wider countryside, which in the UK and other European countries is usually held in private ownership [13]. Groups of landowners and occupiers are incentivised to co-ordinate their actions and to alter land uses, generally away from those uses that would maximise profit for the landholder. Such coordination is not straightforward: neither conservation project managers nor landholders (even if the same) can force cooperation from neighbours [10]. Funds are thus generally required to cover both the direct and opportunity costs of changes in land use and the transactions costs of organising and administering conservation activities. These costs are primarily covered either by government, such as under agri-environment schemes (while the UK belongs to the European Union, predominantly through Pillar 2 of the Common Agricultural Policy), or by funds secured through lotteries, charitable foundations or from private businesses, such as through corporate social responsibility. Conservation organisations also raise funds through membership payments and donations.

The various categories of agents involved in funding ecological restoration are illustrated in Figure 1. The driving force for restoration may rest with the funder (e.g., a private company restoring a mineral extraction site as a requirement of planning), or the implementing organisation (e.g., a conservation NGO such as the Wildlife Trusts or Royal Society for the Protection of Birds). In either case, the funder of such work has a powerful role in shaping the scope and timing of the project. The funder can be thought of as outsourcing specific components of their own programme of activity to external organisations, whether that involves a desire to fund a conservation or restoration project, a direct but more general concern to encourage restoration, or the ability of a restoration project to deliver other aims of the funder (e.g., a restoration project that is funded primarily to provide public access or environmental education on a restoration site).

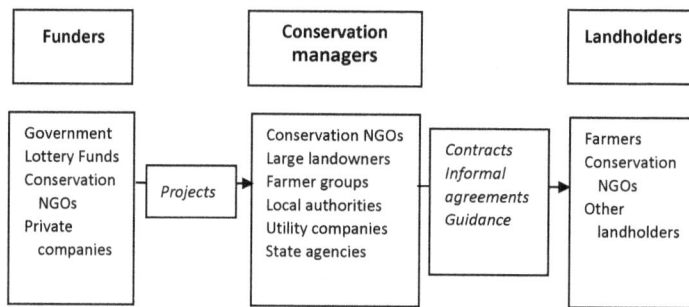

Figure 1. Funding processes for ecological restoration.

Similar approaches have been adopted in other countries. Borgström et al. [14] developed a database of all government funding for ecological restoration in Sweden between 1995 and 2011. They show that funding was predominantly small scale and short-term, reflecting the wider movement towards 'project proliferation'. International development agencies also commonly provide support in developing countries through projects with similar implications. Sayer and Wells [15] have discussed the pathologies of projects implemented for biodiversity conservation in such contexts where local institutions are relatively weak.

The shift towards undertaking activities for public benefit through discrete projects may be seen as emerging from neoliberal initiatives to roll back the state such as in the 'New Public Management' (NPM) [16,17]. The shift towards a neoliberal approach in land conservation initiatives has been witnessed across many countries, such as [18] in the UK, [19] in the USA or [20,21] in Australia. NPM covers a variety of approaches adopted across different countries at different times primarily associated with the rationalistic search for efficiency in public management. Hood [22] suggests seven dimensions that have generally been associated with NPM:

1. A disaggregation of public organizations into separately managed 'corporatized' units for each public sector 'output'.
2. A shift towards greater competition.

3. A move towards management practices used in the private sector.
4. A move towards greater stress on discipline and parsimony in resource use and a search for less costly ways of delivering public services.
5. A move towards 'hands-on management' of public organizations.
6. A move towards more explicit and measurable standards of performance.
7. Attempts to control based on output measures.

This neoliberal approach has important implications for the way in which ecological restoration is undertaken.

3. Methods

A survey was undertaken of managers of large-scale conservation initiatives in the UK [5,6]. These were selected from a database of 800 separate large-scale conservation initiatives compiled by Southampton University. Interviews were undertaken with managers of 27 of them between January and September 2012. Initiatives were chosen purposively, taking into account size, number of landholders, the range of environments across the UK and the nature of the lead organisation. In-depth interviews using a common schedule of questions were conducted face-to-face (n = 23), the rest by phone or *Skype* (n = 4). Questions were open-ended and shared with interviewees before the interview. Questions addressed the objectives and design of the restoration project, the factors influencing its design, and the institutions and practices of partnership working and decision making. Respondents were encouraged to use their own words to clarify and explain complex issues and opportunities were provided to include topics not in the original list of questions. By agreement, all interviews were recorded. They were then transcribed verbatim and coded using *Atlas.ti* to identify and group passages on a common theme. The themes covered the diversity, creation and sustaining of partnerships, formal structures, promoting resilience and community partners. The themes were then linked across interviews. All quotes are anonymous, identified by a unique code.

In the second part of the paper, we explore the potential for drawing out the positive attributes of projectification and adaptive governance into a single approach. Our methodology may be seen as a thought experiment. Reiner and Gilbert [23] characterise a thought experiment as a "design of thought that is intended to test and/or convince others of the validity of a claim". Aligica and Evans [24] argue that it allows "scientists to take existing information that is based on known phenomena and mentally manipulate it into new configurations that can advance the frontiers of knowledge" in circumstances where it is impossible to run experiments in the real world [25]. While best known for their applications in philosophy and physics, thought experiments are also used quite widely in a variety of applications, such as in ecology [26], assessment of the capacity of agricultural land to meet future food demands [27], land requirements for food production [28], and biodiversity conservation [29]. In this context, we seek to integrate positive attributes of projectification and adaptive governance into a single programme for the implementation of ecological restoration. This is clearly not something that can be the subject of experimental testing; rather, we aim to set out an approach based on the experiences discussed in the survey and in the literature. We then identify questions emerging from the analysis that require further research.

4. Projectification in Ecological Restoration

As noted above, whatever the direction of intent, the engagement between the objective of the funder and that of the conservation manager leading the project involves bundling restoration activities into separate work packages, and an element of competitive tendering for the work to be undertaken (in that conservation organisations seeking funding from charities do so competitively, even if they compete with projects that do not have a restoration component). This approach was evident in the responses to our survey. The pressure for projectification was strong:

A lot of other projects, at least the other projects that I'm involved with … tend to literally be 'projects'—they are clearly defined, they've got a short period, maybe 3–4 years of funding, and it's the funding that leaves them … Everything comes together around a funding bid. But if we're talking about large landscape-scale management, the ecosystems, then that's actually not helpful. (Interview 9)

The funder will require a specific output to be delivered, and the project proposal and related contracts will specify the way in which the work will be undertaken, setting out milestones to be achieved to monitor the progress being made towards successful completion. However, in practice, funding is drawn from multiple sources and is not always focused on the delivery of the primary goal of ecological restoration:

We set ourselves, to our funders, we set some targets about how much habitat that was going to be or how much we could change. Now we worked on all sorts of habitat, including grassland and other things, but we were only really reporting on the woodland because that's what the funders paid for. (Interview 2)

Projects are typically funded for a fixed and often relatively short period of time within which the outputs have to be delivered. There will be clauses in the contract specifying actions that will be taken under a range of possible circumstances and these may be invoked if progress fails to match the plan set out in the accepted project proposal. At the end of the agreed contract, if the work is to be continued, a new project may be proposed and a new contract issued, initiating a new project cycle. But funding may not be continued, even if the project has been successful. There may be a need to find new and different objectives in order to attract further project funding. One respondent commented:

With [landscape scale projects], people fund them, and then they say, 'fantastic, it's a great success—we can never fund it again. You've shown what works, now we can't pay for that, you have to do something different'. (Interview 2)

Funds are thus allocated, often on a competitive basis, to projects that can persuade funders that they will deliver the agreed outputs. Calls for applications for funds draw out ideas and proposals for ways in which the funds may be applied in pursuit of the funder's objectives. The process of preparing bids promotes horizontal partnerships amongst stakeholders who can bring different types of resources to address the project objectives. Wolf [30] characterises projects as "temporary platforms for emergent constellations of actors to interact and learn". This offers benefits but also raises challenges:

Each partner is not only able to contribute different skills but also gain something from it … the project partners all have their own strategic plans and their own targets and goals they have to deliver against, and so they hope [the initiative] can help them deliver … It's a sort of happy symbiotic relationship. (Interview 8)

Funders require assurance that projects can deliver the planned outputs with a high degree of financial control. Milestones are erected to facilitate control over the progress of the project and to enable the funder to see whether or not progress is on track to deliver the planned outputs. Funders will evaluate proposals ex ante and rank them, perhaps implicitly, in terms of the ratio of benefits promised over funds sought. Competition amongst applicants will oblige tenderers to reveal their capabilities and costs and allow the funder to select the most promising options [31]. The need to reassure funders creates an incentive for applicants to be relatively unambitious and offer outputs that can be guaranteed. Individual funders have their own requirements.

But there are disadvantages to grant aid, in that you … The grant giving bodies each have their own obsession or expectations or conditions. (Interview S3)

Because the funding may not be specifically targeted to supporting ecological restoration, restoration managers often need to cast around for projects that will be attractive to potential funders.

Sometimes we've got a little bit of money through the National Park, sometimes we've got money through SRDP [The Scottish Rural Development Programme], we're looking at getting money now through landfill to try to develop a project which will help manage the sites and create new habitat, so it really is, you know . . . . (Interview 26)

And there may not be a close match between what funders will support and the aims of the restoration managers:

We've got to work out a system that will bring in different [funding] streams and maybe that's looking at carbon credits or ecosystems services in some way providing something that someone wants to buy into. (Interview 10)

But this still operates on a project-basis.

We are being encouraged to engage with people who are managing our sites to manage them for biodiversity, for carbon, for water, but still within a fairly, not 'ad hoc', but disjointed funding framework which only looks at short-term funding. (Interview 1)

Projects are time-limited so that control over funding can be maintained over time through a series of project cycles: successful projects may be renewed and unsuccessful ones terminated. But restoration processes may be governed by sponsors' own funding limits:

We are at the point of reviewing where we are going next because the Biffa [Biffa Award—funds raised from a landfill tax] money has now run out. (Interview 1)

A failure to secure funds can bring the whole restoration endeavour to an end. One respondent commented *"I mean, if we get this Heritage Lottery Fund my post continues for a period of time as well, but if it doesn't . . . ."* (Interview 25)

This also offers the funder a chance to redistribute funds towards novel outputs or to stakeholders who may be seen to have been underprovided for in previous project cycles, perhaps to enhance perceived spatial or sectoral fairness in fund allocation.

This complexity raises the transactions costs of ecological restoration processes. Many initiatives experience, *"very complex financial administration"* (Interview 8) and organizations often employ a grants officer, or an entire team of people to work on grant applications, especially if the organisation runs a large-scale conservation programme or numerous large-scale conservation areas. However, while restoration is funded on the basis of short-term projects, the emphasis in ecology is towards long-term adaptive management.

5. The Rationale for Adaptive Governance

Projectification involves ceding a degree of control over restoration activities to funders, through agreed funding and outputs. The need to set out intended outcomes to be delivered within a defined period of time encourages an approach that underplays the inherent scientific uncertainty in restoration ecology in practice. Hilderbrand, et al. [32] describe five 'myths' of restoration ecology that may be seen as underpinning the assumptions behind projectification: (1) that we can restore or create an ecosystem that is a carbon copy of a previous or ideal state, (2) that the community and ecosystem assembly process follow a repeatable trajectory, implicitly ignoring uncertainty, (3) that it is possible to accelerate ecosystem development by controlling pathways, such as dispersal, colonization, and community assembly to reduce the time taken to create a functional or desired ecosystem, (4) that we can apply the same restoration techniques in a range of different restoration efforts, and (5) that goals can be achieved by active intervention and unending control or manipulation of physical or biological components of the ecosystem. Morsing et al. [33] examined evidence of the acceptance of these myths in 13 Danish LIFE projects. They found that two assumptions, of a predictable single endpoint and that nature is controllable, were notably frequent in the projects. Schultz et al. [34]

are critical of the European Natura 2000 process for its top down nature and lack of adaptability. "What made sense at the European level and from a biodiversity conservation point-of-view was met by resistance at the local level and by other sectors of society, and there was limited capacity to adapt the process to accommodate their perspectives and solve the conflicts". In a review of natural resource management initiatives in Australia and New Zealand, Curtis et al. [35] describe, as perhaps the most fundamental lesson, a shift from a community-based stance to promote self-reliance to a neoliberal stance of instrumentalising communities to implement higher level strategies.

In response to these uncertainties, ecologists are increasingly advocating adaptive governance, or adaptive co-management, in the restoration of ecosystems following an ecosystem approach. The two terms are often used synonymously [36] and we adopt the former here. Adaptive governance may be defined as "a process by which institutional arrangements and ecological knowledge are tested and revised in a dynamic, ongoing, self-organized process of learning-by-doing" [37]. Ludwig [38] asserts that "the era of management is over", that the management paradigm fails when confronted with complex problems. Chaffin et al. [36] argue that "there is a need, therefore, to champion new approaches to environmental governance capable of confronting landscape-scale problems in a manner both flexible enough to address highly contextualised SESs [social-ecological systems] and dynamic and responsive to adjust to unpredictable feedbacks between social and ecological system components". Schultz et al. [34] make the same argument.

The restoration managers in our survey recognised this approach in their restoration practices:

The way we work together sort of reflects the philosophy of working with natural processes. Natural processes are opportunistic, they aren't always defined, they aren't always very clear ... To a degree we are a bit like that, we sort of react to demands and look at who's got the skills and abilities and time to do it. (Interview 9)

Attempts to restore ecosystems face high levels of uncertainty. The interrelationships and feedbacks amongst ecosystem functions are imperfectly understood and so the consequences of ecosystem interventions are not known in advance. In addition, outcomes are also vulnerable to unpredictable changes in external factors. The consequences of alternative management arrangements thus cannot be predicted with certainty. Ecological restoration involves trade-offs with different actions benefiting different taxa and ecosystem functions, and actions can take decades to become effective [39]. In some circumstances, it is appropriate to follow 'open-ended' approaches to restoration that recognise that long-term ecosystem behaviour involves continual change [40]. Ecosystems can be subject to unpredictable state changes and the risk of this happening is exacerbated by a loss of resilience. It is thus argued that sustainable management should focus on building the resilience of the system [41]. The aim of restoration will often be to build the resilience of the system against unknown future shocks, such as through the maintenance of functional redundancy to underpin service provision, rather than to seek to achieve a predetermined output. Adaptive management [42] recognises this context and argues that management cannot set clear objectives but rather operates on an iterative basis, seeing interventions more as experiments to generate information to feed back into future decisions.

The restoration of ecosystems demands inputs from a broad range of different types of stakeholders. Some will provide land, others will provide entrepreneurship, agricultural management, administrative capacity, funding, voluntary labour, or monitoring and research expertise. These capabilities and resources need to be harnessed and co-ordinated, often relying on high degrees of mutual trust and commitment. Where initiatives involve multiple stakeholders, negotiation strategies are important to resolve differences:

There will, no doubt, continue to be those tensions but we're working them out, certainly, within the partnership. (Interview 27)

Such negotiations will usually involve a mix of private and state organisations in some form of co-management [43,44]. In parallel with the approach to the ecosystem, institutional arrangements also need to be flexible and to adapt and change as experience develops over time. The management of restoration requires the management of both ecological and social systems in an integrated way. Interviewees expressed their belief that community engagement was needed to, "*achieve sustainable land use change*" (Interview 7), and that it was, "*important to keep the focus on maintaining the relationship with local people.*" (Interview 1). Engagement with the local community may add further complexity. One respondent commented:

> *So, some of them I think, genuinely do feel threatened by what we're doing and quite upset by it. But, in terms of the actual landowners, our neighbours, our tenants, I don't think they do feel so threatened.* (Interview 18)

All these processes and relationships take time to develop:

> *You can't build a relationship with people and with a landscape over a year or two years. It takes years of doing that and responding to change as that happens.* (Interview 9)

Taken together, this indicates the potential for the adaptive governance of social-ecological complexity [45,46].

6. Alternative Governance Models

Both approaches, projectification and adaptive governance, have to deal with issues of complexity, multiple stakeholders, the need for leadership, the need for formal commitment and clarity in dealing with conflict resolution. However, the two approaches adopt very different positions as illustrated in Table 1. It might generally be suggested that ecologists would favour adaptive governance while administrators (and arguably politicians) would favour projectification.

Table 1. Comparison of short-term projects and adaptive governance approaches.

	Short-Term Projects	Adaptive Governance
General goal	Efficiency	Resilience
Outputs	Delivery of planned output	Outputs evolve; enhanced knowledge for future management
Monitoring	To check implementation of plan	To better understand system and guide future decisions
Accountability/Power	Control by funder via plan implementation	Shared ownership amongst local community and other stakeholders
Actors	Those bringing planned resources	Those with interests
Knowledge	Implementing contract	Collaborative learning
Management	Minimise deviation from plan	Adaptive approach to new information
Uncertainty	Design plans to minimise impact of uncertainty	Expect and learn from uncertainty
Institutional arrangements	Fixed over project period	Continuous change
Time horizon	Focus on delivery by milestones and projects	Continuity of management and institutional memory over long-term

The strength of projectification in conservation and restoration reflects the influence of New Public Management (NPM) within government, such as in the UK [47]. This has come under increasing criticism in public administration more broadly. Hood and Dixon [48], for instance, have cast doubt on the success of NPM in terms of its capacity to cut the costs of government. Bryson, Crosby and Bloomberg [49] argue that a new movement is emerging to replace NPM that pursues values beyond efficiency and effectiveness. This is variously termed 'public value governance', 'new public governance', or, by Stoker [50], Public Value Management (PVM). Rhodes [51] suggests there may be a return to the 'craft' of public administration. He characterises the new approach as representing a shift from hands-on to hands-off steering by the state, working with and through networks or webs of organisations to achieve shared policy objectives. The criticisms of NPM show clear parallels with the limitations of projectification in ecological restoration. The call for PVM parallels the need for adaptive governance in restoration projects. So, what is the potential for a PVM approach in restoration management?

Stoker [50] sees PVM as a new paradigm for public administration. In contrast to the narrower utilitarian character of NPM, PVM adopts a broader approach to public value which is collectively built through deliberation amongst elected and appointed government officials and key stakeholders.

Governance operates through networks of deliberation and delivery in pursuit of public value. This is fleshed out by four propositions:

- Public interventions are defined by the search for public value.
- There is a need to give more recognition to the legitimacy of a wide range of stakeholders.
- An adaptable and learning-based approach to the challenge of public service delivery is required.

PVM is based on a relational approach to service procurement where client and contractor see each other as partners, looking to sustain a relationship over the long run and not narrowly focussed on any individual contract. PVM emphasises the role of reflection, lesson drawing and continuous adaption. Managers are expected to clarify and express the needs of clients and then tasked with designing and implementing programmes in order to meet them through partnerships. They are tasked with steering networks of deliberation and delivery and with maintaining the overall health of the system. They need to engage in a dialogue in a way that allows for deliberation about choices and alternatives. As contexts and preferences change, this implies a process of continuous evaluation and learning.

This presents significant challenges for efficiency, accountability and equity. The challenge of accountability in particular has been recognised in the context governance networks [34] and of ecological management. Hahn [52] observes that "Governing and ensuring accountability of governance networks, without hampering their flexibility, adaptability, and innovativeness, represents a new challenge for the modern state".

Stoker [50] argues that PVM adopts a different worldview from that of NPM, based on a cooperative perspective: "people need to share and come to endorse each others' viewpoints. The bonds of partnership enable things to get done that no amount of rule setting or incentive providing can deliver". Accountability is achieved by negotiated goal setting and oversight based on complex and continuous exchange among leadership and checks and balances to that leadership to ensure that leadership is facilitative. Accountability then arises from more extended citizen involvement. It thus tends to be informal rather than formal. Romzek and LeRoux [53] and Romzek et al. [54] have studied informal accountability amongst social service networks in the United States. They identify the development of social norms and facilitative behaviour in order to maintain order and accountability amongst collaborating agencies and sub-contractors with complementary but different missions, agendas and protocols. These are supported by an informal system of rewards and sanctions and relationship building but are threatened by organisational obstacles. Informal accountability "emerges from the unofficial expectations and discretionary behaviours that take shape through repeated interactions among network members cognizant of their interdependence in pursuit of their shared goal(s)" [54]. It can be challenged by financial pressures that undercut collaborative activities such as relationship building.

Butler et al. [55] have made a similar observation in the context of collaborative implementation of ecological restoration on US forest land. They comment [55] that the process "appears to strengthen USFS [United States Forest Service] accountability to collaborators through such informal and relational mechanisms where understandings and concerns emerge through collaborative interaction" and that "Multiparty monitoring, provides a direct set of mechanisms for strengthening accountability as stakeholder values and perspectives are integrated into implementation processes through participation and dialogue". Hahn [52], in an analysis of the multilevel governance network of Kristianstads Vattenrike Biosphere Reserve in Southern Sweden, refers to shared accountability in this type of context.

It is accepted that ensuring accountability is not straightforward in that it requires high levels of trust and active citizen engagement, raising fundamental questions about the nature of democracy. There is an inherent tension between management and democracy. "Vigilance and regular critical review by all the partners in the system is central to ensuring that the promise of both stakeholder democracy and management is delivered" [51]. Rhodes suggests that, put simply, "management and markets are the priority for NPM while delivering services to citizens is the priority for New Public Governance" or PVM. To extend this simplification, we might suggest that ecological restoration is being managed on the basis of the former while in many ways better fitting into the latter. But as Rhodes further emphasises, what is important is to identify what works and what skills are required in a particular context [51].

7. Towards an Architecture of Funding for Ecological Restoration

In this section, as a form of thought experiment, we seek to draw together the strengths of projectification and adaptive governance into a single process. Notwithstanding the criticisms, projectification has addressed a variety of the challenges faced by government and other funders in providing support for ecological restoration in the UK. It stimulates ideas and proposals for restoration from a range of stakeholders, it promotes collaboration amongst stakeholders who can bring different resources and capabilities to the project (at least in the short-term), competitive bidding creates incentives for organisations to leverage matched funding and other resources and promotes cost-effectiveness, it provides for financial control and accountability as well as control over the management of ecosystem interventions. It ensures regular opportunities to review objectives, check on progress and reallocate resources to other, potentially more effective groups and projects over time. But there are significant limitations too. Projectification encourages, and in some instances, requires conservationists to select less relevant, unambitious and potentially counterproductive objectives for their activities. It sets short time horizons over which projects need to be able to show demonstrable outputs, it interrupts longer-term efforts to build relationships and trust amongst stakeholders, it raises transactions costs in terms of the resources required to prepare proposals and bid, often unsuccessfully, for funds and report on completed projects, it fails to ensure continuity of employment for those engaged in conservation activities.

Adaptive governance potentially offers solutions to many of these limitations, but at the risk of undermining the benefits. Plummer et al. [56] have undertaken a systematic review of the literature linking adaptive co-management with environmental governance. Olsson et al. [37] identify seven features that support the emergence of adaptive co-management of social-ecological systems:

- Enabling legislation that creates social space for ecosystem management,
- Funds for responding to environmental change and for remedial action,
- Ability to monitor and respond to environmental feedbacks,
- Information flow and social networks for ecosystem management,
- Combination of various sources of information for ecosystem management,
- Sense-making for ecosystem management,
- Arenas of collaborative learning for ecosystem management.

They conclude "The shared vision of the actors and the self- organizing process, supported and framed by enabling legislation and governmental institutions, have the potential to expand desirable stability domains of a region. It creates an 'adaptive dance' between resilience and change with the potential to sustain complex social–ecological systems."

The adoption of adaptive approaches is not straightforward. Westgate et al. [42] comment that adaptive management has rarely been achieved in practice. Allan and Curtis [56] could also find few examples of adaptive management in use and concluded that the deeply embedded culture of resource management that requires managers to demonstrate attainment through the achievement of milestones and targets to ensure continued funding thwarted any opportunity for the collaborative and holistic thinking, learning and experimentation required for adaptive management. Adaptive management is often seen as too open-ended for rigorous financial control and too uncertain in terms of planned outputs. The question is whether there is some alternative process that can capture the advantages of both approaches, drawing on the emerging shift towards PVM in public administration.

Any system to promote ecosystem restoration needs to promote entrepreneurial activity, to enable administration and facilitation, and to provide funding to cover these functions and to cover the direct and opportunity costs of changes in land management. It may well be that in some circumstances, the restoration of ecosystems can be supported by a new funding stream through a Payment for Ecosystem Services (PES) scheme [57,58]. This option needs to be explored, subject to the primary restoration mission. The goals should not be conceived narrowly in terms of biodiversity conservation.

As we have stressed, ecological restoration demands a long-term commitment to consistent land management, and so long-term funding must be potentially available. There needs to be an assured fund dedicated to support restoration activities in the long-term. But funding for individual restoration initiatives or for specific partnerships cannot be unconditional. The funder needs assurance that the approach is in some sense 'on track' and that the management is cost-effective. However, it may not be possible to identify in advance discrete milestones and specific long-term outcomes to be achieved. In place of this, there needs to be periodic, transparent ex post deliberation as to the quality and direction of management. This allows greater discretion for the use of funds in ecosystem restoration, subject to its subsequent justification. There is a parallel here with the governance of charities more generally, but in addition, it will require assessment by experienced individuals who can review the quality of ecosystem management and the effectiveness of the expenditure committed in the previous period.

Rather than focussing on individual projects one at a time, funding for ecological restoration might be thought of as developing funding programmes, setting a series of time limited projects within the context of a longer-term programmatic framework. This funding process might itself adopt an adaptive approach where individual projects are seen as experiments within the context ecological restoration funding.

This suggests the need for a core public funding source that is dedicated to the purpose of ecological restoration. This might be regionally based and should have the:

- Capability to assess restoration priorities within its locality of responsibility but across different ecosystem services,
- Ability to fund direct, opportunity and transactions costs incurred in undertaking restoration work,
- Potential to deliver long-term continuous funding where justified,
- Capability to assess progress in adaptive governance.

The core funder would invite potential restoration managers to tender for ecological restoration projects. Successful projects would be funded for a fixed period of time in order to initiate restoration activities, covering the costs of the identification of ecological priorities and planned ecological interventions, liaison and building relationships with relevant stakeholders, and identification of alternative potential funding sources that would be consistent with and complement the ecological priorities. Agri-environment schemes represent a major potential funding source to support land managers. There would also be potential for the development of Payment for Ecosystem Services schemes, commercial activities and sponsorship from other public or private sources. This could lead to plans for ecological restoration in the longer term, recognising potential complementarities in the delivery of different ecosystem services to different groups of beneficiaries.

If successful, this would allow funding to be offered for a programme (national or regional) of sequential, time-limited projects to implement an adaptive approach to ecological restoration. It is to be expected that the core funding provided by the core funder would enable the restoration manager to leverage additional funding streams, for activities, potentially supported by other sponsors but complementing the aims of the ecological restoration programme. The activities involved in the primary ecological restoration and in the wider programme of ecosystem service delivery would be assessed periodically by the core funder and funding would be continued where progress was judged to be satisfactory. Adaptive governance raises challenges for conventional approaches to evaluation, requiring analysis that goes beyond the physical impact on the ecosystem [59] to give attention to the ecological and economic components as well as the process component that looks at the role of institutions and power [56]. Restoration should be viewed in its social and political context [60]. Lockwood et al. [61] propose seven governance principles that offer a potential basis against which assessment might be conducted. The assessment would consider progress in ecological land management, the development in the delivery of ecosystem services, the changes in the assessed resilience of the ecosystem, and the development and quality of networks amongst stakeholders engaged in the various aspects in support of ecological restoration.

The approach is illustrated schematically in Figure 2. This represents core funding provided by a core public sector funder for up to nine projects over four sequential periods. Five projects are initially funded in period 1. At the end of the period, four are considered to be making satisfactory progress, while one is terminated. Three of the other projects are able to reduce their reliance on core funding by identifying alternative sources, such as through Payment for Ecosystem Services, commercial enterprises or donations. This allows funding to be awarded to two further projects. This process continues so that nine projects are receiving funds from the core funder in period 4. It may be expected that the reliance on core funding varies between projects. Thus, for example, project 7 receives full funding over three periods in a context where there are no options for funding from alternative sources.

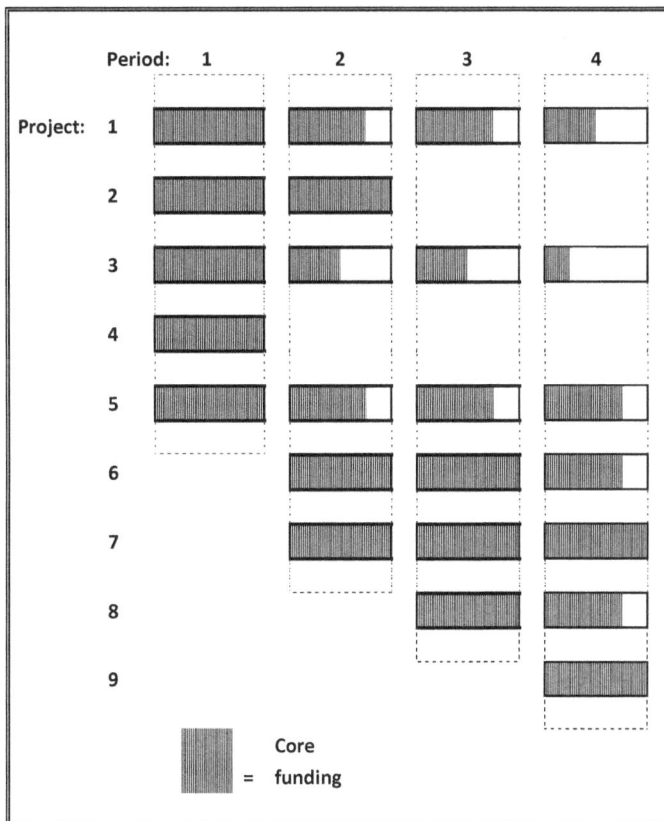

Figure 2. Illustration of a funding programme over four periods.

Under this approach, accountability could be attained in various ways. There would be a direct vertical accountability from the core public sector funder, itself under democratic control, to maintain oversight over the activities of the ecological restoration manager. The longer relationship between funder and project managers allows them to build trust in each other's role. Failure to meet the required standards could lead to the termination of the programme. Projects funded by other funders would be subject to scrutiny by their funders and the delivery of PES schemes would be subject to scrutiny by the beneficiaries of the ecosystem services themselves. Over time, as a network of stakeholders develops and interactions amongst stakeholders become more complex, then informal accountability will become more important. The partners in restoration activities will look to each other for transparency and cost effectiveness in expenditure, potentially supporting this through informal sanctions. Failure of one partner potentially undermines the opportunities of the others and so they have an incentive to monitor each other. There are parallels here with analysis of self enforcement in collective action (cf. Ostrom [62]).

The funding programme could be judged to be concluded when the governance of the ecosystem becomes self-supporting through internal decision-making processes, informal and internal accountability and external sources of funding in payment for benefits provided to external stakeholders. This is not to suggest that the problem is 'solved' but rather that the institutions to manage the system have developed sufficiently to become self-sustaining. However, this may potentially never be achieved while the ecosystem is delivering public goods to beneficiaries outside of the local area under ecological management. In this context, public funding would continue to the extent that the ecosystem management is delivering public good benefits for which beneficiaries are not making a direct contribution.

This identifies a number of questions that deserve further research:

- How to prioritise restoration options, especially in the context of climate change and when the outcomes of restoration actions may be uncertain. Funding needs to be allocated to projects that are regarded as being most important in some sense. A core funder would need to establish systematic and clear criteria for judging applications for funding.
- How to evaluate open-ended adaptive projects as they are undertaken. Conventional project evaluation concentrates on the extent to which projects have delivered their planned outputs and, ideally, outcomes. In the context of adaptive government, more attention will need to be given to the processes followed and the outcomes that might not have been anticipated at the start of the project.
- The merits of alternative governance structures for restoration programmes. Ecological restoration projects involve a range of stakeholders bringing different resources and capabilities and taking different responsibilities and risks. We need to learn more from the experience with projects to date and to explore the implications of alternative arrangements [63].
- The roles of formal and informal accountability. Ecological restoration projects are largely located in the non-profit sector. They are not under direct democratic control as can be the case in the public sector, nor under pressure to meet constraints imposed by markets, as in the private sector. However, their role is to operate in the social interest. We have suggested that there may be scope for a shift from formal to informal methods of maintaining accountability, but this needs to be explored further and tested.

8. Conclusions

Conservation effort is increasingly being directed towards the management and restoration of larger areas. Managers of ecological restoration initiatives are under competing pressures illustrated in our survey of large-scale conservation initiatives. Funding sources demand short-term projects to be won competitively with discrete and well-defined outputs. But at the same time, ecologists are questioning this approach, arguing that restoration should adopt adaptive governance that brings together groups of stakeholders undertaking ecological interventions on an experimental basis and accumulating information towards enhanced ecosystem resilience. There are parallel developments in thinking between ecology and public administration where the neoliberal approach embodied in NPM is being challenged by a broader conception of value, working though networks of partners, such as represented by PVM.

Both approaches offer particular advantages, suggesting that the issue is not one of simply selecting one or other approach. A key issue concerns accountability. Adaptive governance advocates a long-term perspective and accepts uncertain outcomes. But in this context, how can funders be assured that funds will be accounted for and 'well' spent? In this context, we have sought to sketch an institutional architecture that draws the two approaches together. This centres around a core public fund that is dedicated to ecological restoration administered by or on behalf of government. It supports programmes of long-term funding for ecological restoration, subject to evidence that the activities and progress meet certain standards. These standards relate to institutional development, network creation, building trust and ecological understanding, and levering additional funding, as much as they do to the achievement of predetermined environmental outcomes.

We argue that this approach has the potential to offer greater continuity for ecological management, with a clearer focus on ecosystem functions and lower levels of transaction costs. But at the same time, it creates an incentive for restoration managers to seek and build partnership arrangements and to identify the beneficiaries of ecosystem services who may be expected to pay for the delivery of a service. It will take time to develop trust amongst potential partners and to identify and introduce institutional arrangements under which payment from beneficiaries will be forthcoming. There is thus an argument for public funding in the short-term while new institutions are developed and implemented. At the same time, ecological restoration can also generate public goods that, given the level of transactions costs, will never be supported through market processes. Longer-term public funding is legitimated by these missing markets.

There are thus various possible sources of funding beyond direct subventions from taxpayers. We argue that ecological restoration should be seen in a broader context of ecosystem governance [63], alongside management of land and water [64]. It should also be viewed in the context of agricultural policy where agri-environment payments [65] as well as direct payments made under Pillar 1 of the Common Agricultural Policy, or a subsequent policy, can be directed towards forms of land use and land management that make a positive contribution towards the delivery and maintenance of ecosystem services. The exit of the UK from the European Union offers an opportunity in the UK to explore a wider range of policy approaches.

More research is needed in order to explore the conceptual frameworks and the practicalities of alternative forms of ecosystem governance. But the arguments here about the need for flexibility and an adaptive approach in restoration projects are already clear to project managers.

We say we have an adaptive management technique and I think that's probably the best way to describe it. Try it, if it doesn't work, change the rules. (Interview 25)

Managers also recognise that adaptive management demands flexibility in project funding,

If we haven't delivered, you know, we don't stand in particularly good position in terms of the grants and so on, so we're starting off modestly, we're seeing whether or not it works and if does then we'll try and do more of it. (Interview 22)

More needs to be done to synthesise experience and mainstream successful institutional models for ecological restoration. However, given the spatial heterogeneity in the physical and social environments there will be no universal solution. Institutional models will still need to be tailored to individual local contexts, as successful project managers know only too well.

Acknowledgments: Funding for the initial surveys of large-scale conservation areas was provided by Natural England, Scottish Natural Heritage and the University of Cambridge Moran Fund. However, they bear no responsibility for the contents of this paper.

Author Contributions: William M. Adams led on the design and implementation of the survey. Ian Hodge wrote the paper with William M. Adams.

Conflicts of Interest: The authors declare no conflict of interest.

References

1. Lawton, J.H.; Brotherton, P.N.M.; Brown, V.K.; Elphick, C.; Fitter, A.H.; Forshaw, J.; Haddow, R.W.; Hilborne, S.; Leafe, R.N.; Mace, G.M.; et al. *Making Space for Nature: A Review of England's Wildlife Site and Ecological Network*; Department for Environment, Food and Rural Development: London, UK, 2010.
2. Adams, W.M.; Hodge, I.D.; Sandbrook, L. New spaces for nature: The re-territorialization of biodiversity conservation under neoliberalism in the UK. *Trans. Inst. Br. Geogr.* **2014**, *39*, 574–588. [CrossRef]
3. Crooks, K.R.; Sanjayan, M. *Connectivity Conservation*; Cambridge University Press: Cambridge, UK, 2006.
4. Fitzsimons, J.; Pulsford, I.; Wescott, G. *Linking Australia's Landscapes: Lessons and Opportunities for Large-Scale Conservation Networks*; CSIRO Publishing: Melbourne, Australia, 2013.
5. Macgregor, N.A.; Adams, W.M.; Hill, C.T.; Eigenbrod, F.; Osborne, P.E. Large-scale conservation in Britain. *Ecos Rev. Conserv.* **2012**, *33*, 13–23.
6. Adams, W.M.; Hodge, I.D.; Macgregor, N.A.; Sandbrook, L. Creating restoration landscapes: partnerships in large-scale conservation in the UK. *Ecol. Soc.* **2016**. [CrossRef]
7. Sjöblom, S.; Löfgren, K.; Godenhjelm, S. Projectified politics—temporary organisations in a public context. *Scand. J. Public Adm.* **2013**, *17*, 3–12.
8. Sjöblom, S. Administrative short-termism—A non-issue in environmental and regional governance. *J. Environ. Policy Plan.* **2009**, *11*, 165–168. [CrossRef]
9. Hughes, F.M.R.; Adams, W.M.; Butchart, S.H.M.; Field, R.H.; Peh, K.S.-H.; Warrington, S. The challenges of integrating biodiversity and ecosystem services monitoring and evaluation at a landscape-scale wetland restoration project in the UK. *Ecol. Soc.* **2016**, *21*, 10. [CrossRef]
10. Martin, P. Ecological restoration of rural landscapes: Stewardship, governance, and fairness. *Restor. Ecol.* **2016**, *24*, 680–685. [CrossRef]
11. Sjöblom, S.; Andersson, K.; Marsden, T.; Skerratt, S. *Sustainability and Short-Term Policies: Improving Governance in Spatial Policy Interventions*; Ashgate: Farnham, UK, 2012.
12. Sjöblom, S.; Godenhjelm, S. Project proliferation and governance-implications for environmental management. *J. Environ. Policy Plan.* **2009**, *11*, 169–185. [CrossRef]
13. Hodge, I.; Hauck, J.; Bonn, A. The alignment of agricultural and nature conservation policies in the European Union. *Conserv. Biol.* **2015**, *29*, 996–1005. [CrossRef] [PubMed]
14. Borgström, S.; Zachrisson, A.; Eckerberg, K. Funding ecological restoration policy in practice-patterns of short-termism and regional biases. *Land Use Policy* **2016**, *52*, 439–453. [CrossRef]
15. Sayer, J.; Wells, M.P. The pathology of projects. In *Getting Biodiversity Projects to Work: Towards More Effective Conservation and Development*; McShane, T.O., Wells, M.P., Eds.; Columbia University Press: New York, NY, USA, 2004; pp. 35–48.
16. Hood, C. A public management for all seasons. *Public Adm.* **1991**, *69*, 3–19. [CrossRef]
17. Osborne, D.; Gaebler, T. *Reinventing Government: How the Entrepreneurial Spirit is Transforming the Public Sector*; Plume: New York, NY, USA, 1992.
18. Hodge, I.D.; Adams, W.M. Neoliberalization, rural land trusts and institutional blending. *Geoforum* **2012**, *43*, 472–482. [CrossRef]
19. Robertson, M. The neoliberalization of ecosystem services: Wetland mitigation banking and problems in environmental governance. *Geoforum* **2004**, *35*, 361–373. [CrossRef]
20. Lockie, S.; Higgins, V. Roll-out neoliberalism and hybrid practices or regulation in Australian agri-environmental governance. *J. Rural Stud.* **2007**, *23*, 1–11. [CrossRef]
21. Race, D.H.; Curtis, A. Reflections on the effectiveness of market-based instruments to secure long-term environmental gains in South East Australia. *Soc. Nat. Resour.* **2013**, *26*, 1050–1065. [CrossRef]
22. Hood, C. The 'New Public Management' in the 1980s: Variations on a theme. *Account. Organ. Soc.* **1995**, *20*, 93–109. [CrossRef]
23. Reiner, M.; Gilbert, J. Epistemological resources for thought experimentation in science learning. *Int. J. Sci. Educ.* **2000**, *22*, 489–506. [CrossRef]
24. Aligica, P.D.; Evans, A.J. Thought experiments, counterfactuals and comparative analysis. *Rev. Aust. Econ.* **2009**, *22*, 225–239. [CrossRef]
25. Brown, J.R.; Fehige, Y. Thought Experiments. Available online: http://plato.stanford.edu/entries/thought-experiment/ (accessed on 18 September 2016).

26. Wilkinson, D.M. The fundamental processes in ecology: A thought experiment on extraterrestrial biospheres. *Biol. Rev.* **2003**. [CrossRef]
27. Koh, L.P.; Koellner, T.; Ghazoul, J. Transformative optimisation of agricultural land use to meet future food demands. *PeerJ* **2013**. [CrossRef] [PubMed]
28. Kastner, T.; Nonhebel, S. Changes in land requirements for food in the Philippines: A historical analysis. *Land Use Policy* **2010**, *27*, 853–863. [CrossRef]
29. Hoffmann, M.; Duckworth, J.W.; Holmes, K.; Mallon, D.P.; Rodrigues, A.S.; Stuart, S.N. The difference conservation makes to extinction risk of the world's ungulates. *Conserv. Biol.* **2015**, *29*, 1303–1313. [CrossRef] [PubMed]
30. Wolf, S. Temporal dimensions of governance: A critical analysis of projects. In *Sustainability and Short-Term Policies*; Sjöblom, S., Ed.; Ashgate: Farnham, UK, 2012; pp. 181–199.
31. Whitten, S.M.; Reeson, A.; Windle, J.; Rolfe, J. Designing conservation tenders to support landholder participation: A framework and case study assessment. *Ecosys. Serv.* **2013**, *6*, 82–92. [CrossRef]
32. Hilderbrand, R.H.; Watts, A.C.; Randle, A.M. The myths of restoration ecology. *Ecol. Soc.* **2005**, *10*, 19.
33. Morsing, J.; Frandsen, S.; Vejre, H.; Raulund-Rasmussen, K. Do the principles of ecological restoration cover EU LIFE Nature co-funded projects in Denmark? *Ecol. Soc.* **2013**. [CrossRef]
34. Schultz, L.; Folke, C.; Osterblom, H.; Olsson, P. Adaptive governance, ecosystem management, and natural capital. *Proc. Natl. Acad. Sci. USA* **2015**, *112*, 7369–7374. [CrossRef] [PubMed]
35. Curtis, A.; Ross, H.; Marshall, G.R.; Baldwin, C.; Cavaye, J.; Freeman, C.; Carr, A.; Syme, G.J. The great experiment with devolved NRM governance: lessons from community engagement in Australia and New Zealand since the 1980s. *Aust. J. Environ. Manag.* **2014**, *21*, 175–199. [CrossRef]
36. Chaffin, B.C.; Gosnell, H.; Cosens, B.A. A decade of adaptive governance scholarship: synthesis and future directions. *Ecol. Soc.* **2014**. [CrossRef]
37. Olsson, P.; Folke, C.; Berkes, F. Adaptive co-management for building resilience in social-ecological systems. *Environ. Manag.* **2004**, *34*, 75–90. [CrossRef] [PubMed]
38. Ludwig, D. The era of management is over. *Ecosystems* **2001**, *4*, 758–764. [CrossRef]
39. Oliver, T.H.; Isaac, N.J.B.; August, T.A.; Woodcock, B.A.; Roy, D.B.; Bullock, J.M. Declining resilience of ecosystem functions under biodiversity loss. *Nat. Commun.* **2015**. [CrossRef] [PubMed]
40. Hughes, F.M.R.; Stroh, P.; Adams, W.M. When is open-endedness desirable in restoration projects? *Restor. Ecol.* **2012**, *20*, 291–295. [CrossRef]
41. Scheffer, M.; Carpenter, S.R.; Foley, J.; Folke, C.; Walker, B. Catastrophic shifts in ecosystems. *Nature* **2001**, *413*, 591–596. [CrossRef] [PubMed]
42. Westgate, M.J.; Likens, G.E.; Lindenmayer, D.B. Adaptive management of biological systems: A review. *Biol. Conserv.* **2013**, *158*, 128–139. [CrossRef]
43. Carlsson, L.; Berkes, F. Co-management: Concepts and methodological implications. *J. Environ. Manag.* **2005**, *75*, 65–76. [CrossRef] [PubMed]
44. Berkes, F. Evolution of co-management: Role of knowledge generation, bridging organizations and social learning. *J. Environ. Manag.* **2009**, *90*, 1692–1702. [CrossRef] [PubMed]
45. Folke, C.; Hahn, T.; Olsson, P.; Norberg, J. Adaptive governance of social-ecological systems. *Annu. Rev. Environ. Resour.* **2005**, *30*, 441–473. [CrossRef]
46. Armitage, D.R.; Plummer, R.; Berkes, F.; Arthur, R.I.; Charles, A.T.; Davidson-Hunt, I.J.; Diduck, A.P.; Doubleday, N.C.; Johnson, D.S.; Marschke, M.; et al. Adaptive co-management for social-ecological complexity. *Front. Ecol. Environ.* **2008**, *7*, 95–102. [CrossRef]
47. Levy, R. New Public Management end of an era? *Public Policy Adm.* **2010**, *25*, 234–240. [CrossRef]
48. Hood, C.; Dixon, R. A model of cost-cutting in government? The great management revolution in UK central government reconsidered. *Public Adm.* **2013**, *91*, 114–134. [CrossRef]
49. Bryson, J.M.; Crosby, B.C.; Bloomberg, L. Public value governance: moving beyond traditional public administration and the new public management. *Public Adm. Rev.* **2014**, *74*, 445–456. [CrossRef]
50. Stoker, G. Public value management: A new narrative for networked governance? *Am. Rev. Public Adm.* **2006**, *36*, 41–57. [CrossRef]
51. Rhodes, R.A.W. Recovering the craft of public administration. *Public Adm. Rev.* **2015**. [CrossRef]
52. Hahn, T. Self-organized governance networks for ecosystem management: Who is accountable? *Ecol. Soc.* **2011**, *16*, 18.

53. Romzek, B.S.; LeRoux, K. A preliminary theory of informal accountability amongst network organizational actors. *Public Adm. Rev.* **2012**, *72*, 442–453. [CrossRef]

54. Romzek, B.; LeRoux, K.; Johnston, J.; Kempf, R.J.; Piatak, J.S. Informal accountability in multisector service delivery collaborations. *J. Public Adm. Res. Theory* **2014**, *24*, 813–842. [CrossRef]

55. Butler, W.H.; Monroe, A.; McCaffrey, S. Collaborative implementation for ecological restoration on US public lands: Implications for legal context, accountability, and adaptive management. *Environ. Manag.* **2015**, *55*, 564–577. [CrossRef] [PubMed]

56. Plummer, R.; Armitage, D.R.; de Loe, R.C. Adaptive governance and its relationship to environmental governance. *Ecol. Soc.* **2013**. [CrossRef]

57. Engel, S.; Pagiola, S.; Wunder, S. Designing payments for environmental services in theory and practice: An overview of the issues. *Ecol. Econ.* **2008**, *65*, 663–674. [CrossRef]

58. Vatn, A. An institutional analysis of payments for environmental services. *Ecol. Econ.* **2010**, *69*, 1245–1252. [CrossRef]

59. Wortley, L.; Hero, J.-M.; Howes, M. Evaluating ecological restoration success: A review of the literature. *Restor. Ecol.* **2013**, *21*, 537–543. [CrossRef]

60. Baker, S.; Eckerberg, K. A policy analysis perspective on ecological restoration. *Ecol. Soc.* **2013**, *18*, 17. [CrossRef]

61. Lockwood, M. Good governance for terrestrial protected areas: A framework, principles and performance outcomes. *J. Environ. Manag.* **2010**, *91*, 754–766. [CrossRef] [PubMed]

62. Ostrom, E. *Understanding Institutional Diversity*; Princeton University Press: Princeton, NJ, USA, 2005.

63. Hodge, I. *The Governance of the Countryside*; Cambridge University Press: Cambridge, UK, 2016.

64. Helm, D. Catchment Management, Abstraction and Flooding: The Case for a Catchment System Operator and Coordinated Competition. Available online: http://www.dieterhelm.co.uk/assets/secure/documents/Catchment-Management-Abstraction-and-Flooding.pdf (accessed on 1 August 2016).

65. Plieninger, T.; Schleyer, C.; Schaich, H.; Ohnesorge, B.; Gerdes, H.; Hernández-Morcillo, M.; Bieling, C. Mainstreaming ecosystem services through reformed European agricultural policies. *Conserv. Lett.* **2012**, *5*, 281–288. [CrossRef]

Article

The Community-Conservation Conundrum: Is Citizen Science the Answer?

Mel Galbraith [1,2,*], Barbara Bollard-Breen [2] and David R. Towns [2,3]

[1] Environmental and Animal Sciences, Unitec Institute of Technology, Private Bag 92025, Auckland 1142, New Zealand

[2] Institute for Applied Ecology New Zealand, Auckland University of Technology, Private Bag 92006, Auckland 1142, New Zealand; bbreen@aut.ac.nz (B.B.-B.); dtowns@aut.ac.nz (D.R.T.)

[3] New Zealand Department of Conservation, Private Bag 68908, Auckland 1145, New Zealand

* Correspondence: mgalbraith@unitec.ac.nz; Tel.: +64-2-73879720

Academic Editors: Jeffrey Sayer and Chris Margules

Received: 8 August 2016; Accepted: 25 October 2016; Published: 31 October 2016

Abstract: Public participation theory assumes that empowering communities leads to enduring support for new initiatives. The New Zealand Biodiversity Strategy, approved in 2000, embraces this assumption and includes goals for community involvement in resolving threats to native flora and fauna. Over the last 20 years, community-based ecological restoration groups have proliferated, with between 600 and 4000 identified. Many of these groups control invasive mammals, and often include protection of native species and species reintroductions as goals. Such activities involve the groups in "wicked" problems with uncertain biological and social outcomes, plus technical challenges for implementing and measuring results. The solution might be to develop a citizen science approach, although this requires institutional support. We conducted a web-based audit of 50 community groups participating in ecological restoration projects in northern New Zealand. We found great variation in the quality of information provided by the groups, with none identifying strategic milestones and progress towards them. We concluded that, at best, many group members are accidental scientists rather than citizen scientists. Furthermore, the way community efforts are reflected in biodiversity responses is often unclear. The situation may be improved with a new approach to data gathering, training, and analyses.

Keywords: ecological restoration; citizen science; monitoring; conservation volunteering; New Zealand; wicked problems

1. Introduction

Public participation theory is the direct or indirect involvement of concerned stakeholders in policies, plans, or programs in which they have an interest [1,2]. It is assumed that public participation is beneficial to society in that empowering communities leads to enduring support for new initiatives, that those affected by decisions or actions should be involved in their implementation, and that communities that work together can achieve outcomes that are broader than those that can be achieved by individuals alone [1,3–5]. However, the kinds of decisions that might be appropriate, what is involved in participation and how best to implement it, are often unclear (e.g., [6]).

Conservation projects are regarded by many authors [4,7,8] as ideally suited to participation activities because they provide access to local knowledge, sustainable outcomes through on-going motivation, building of capacity through the acquisition of transferable skills, sharing of responsibilities, acceleration of change through growth of education, awareness and trust, economies of scale, and less costly enforcement through self-regulation [4,9]. Building on these advantages, community-based voluntary participation in ecological restoration (conservation) has shown global growth [10–12].

The motivations for voluntary participation in ecological restoration projects are well studied (e.g., [10,13–15]), and may be summarised as helping the environment, learning more about the environment (ecological literacy), social belonging (social networks), and personal growth. Although government agencies may be legally bound to restore degraded environments, many have taken advantage of this growth of interest in public participation, and have included increased engagement of communities in conservation as performance indicators (e.g., New Zealand Department of Conservation, DOC) [16].

The concept of public participation is implicit in the New Zealand Biodiversity Strategy (NZBS) [4,17] adopted by the New Zealand Government in 2000. The Strategy acknowledges that lack of understanding and awareness of biodiversity is a barrier to biodiversity conservation, and identifies the engagement of communities and individuals to conserve and enhance New Zealand's biodiversity as a potential tool. Citizen science, the engagement of non-professionals in scientific investigations [18], is not specifically mentioned in the NZBS, but the concept is certainly implied. For example, Goal 1 of the strategy ([17], p. 15) refers to communities having to share responsibility "equitably" for the conservation of New Zealand's biodiversity. This encompasses a breadth of potential outcomes of citizen science, including the gain of scientific literacy, fostering and strengthening of relationships between citizens and professional agencies, use of knowledge for advocacy, and to influence political decisions [19,20]. The action plan for Goal 1 advocates community involvement, using "participatory projects" ([17], p. 102) as a tool to resolve the threats to native flora and fauna, and to promote the sustainable use of natural resources. Over the last 20 years, community-based ecological restoration groups have proliferated in New Zealand [14,21], with more than 3500 ecological restoration projects [22] involving an estimated 4000 community groups [23–25].

There is general agreement that there are many levels of public participation. For conservation problems, a form of participation called adaptive co-management is advocated to address the socio-ecological complexities of environmental problems [26]. Here, the scale of the partnerships formed, the participants and type of arrangement formed, should reflect the complexity of problems being addressed. The complex nature of conservation problems is evident in New Zealand, where ecological restoration, especially on islands, almost always involves management of invasive species [17,27], and the translocation of species to refugia for conservation gains or to fill taxonomic gaps [28,29]. Both of these activities, increasingly undertaken by community groups, require numerous sequences of complex actions and decisions.

Invasive plants can take some time to naturalise [30], and may require an equally lengthy time to eradicate or control [31], but invasive mammalian predators, even in low densities, can have an immediate and drastic impact on native biota [32,33]. For this reason, many restoration projects in New Zealand prioritise mammal control, even though such strategies seldom have definite end-points unless eradication is achievable [34] and even then, reinvasion may remain a perpetual risk [35]. If community groups aim to deal with introduced predators, they are confronted by significant and complex hurdles. The groups may face systematic removal (or control) of multiple species, which requires planning, logistics, technical requirements, and funds—elements that are acknowledged as a challenge for professional managers [36], and are potentially beyond most community groups working independently of other stakeholders [37,38]. For example, the eradication of rats from large islands involves the aerial spread of rodenticide using helicopters equipped with sophisticated Global Positioning Systems [39]. In New Zealand, there are restrictions on who can use the rodenticide, and many safety requirements must be met while the products are loaded and spread. There are also regulations about the discharge of toxins into the air and water that usually require resource consent from local authorities.

Like the management of invasive species, translocations of native species involve an array of complex hurdles for practitioners to address [40,41]. Translocation proposals must provide the rationale and justification for translocation, and consider logistics, viability of both source and transferred populations, habitat requirements, welfare needs during transfer, disease screening needs, and funding [42]. Increasingly, genetic issues need to be addressed as each translocation event is, potentially, a genetic bottleneck [43]. Furthermore, extended post-release monitoring of both source and transferred populations is often a required component of translocation events [44,45]. Community involvement in species translocations as part of the restoration process, whether wholly community-led or joint community/agency initiatives, is increasing [24,46]. For approved species translocations in New Zealand during the period of 2002–2012, community participation increased from 16% to 71% [46]. Community groups are unlikely to be fully aware of the complexity of undertaking translocations [47], although community participation in translocations is now considered to be an essential component of conservation advocacy [29,48].

These restoration activities have inherent, and potentially unforeseen challenges, which are typical of conservation management worldwide. For example, a fundamental part of halting the biodiversity declines identified in the NZBS is, of necessity, ecological restoration that includes pest eradication [17]. In a study of ecological restoration groups in New Zealand [24], at least 75% were involved in animal pest control. Should communities embark on ambitious habitat restoration projects that involve killing unwanted organisms, they will lurch unsuspectingly into the realm of "wicked" problems [49,50], which are those with complex and interconnected components, uncertain biological and social outcomes, which may operate over short or long time-scales, have ambiguous definition of scope and boundaries, and be subject to controversy and locally variable social constraints [38,51,52].

Reviews of the implementation of the NZBS [25,53,54] acknowledge the increase in community-sourced participants in conservation activities. Although the NZBS urges agencies to ensure that individuals and communities have the knowledge and technical skills to participate in biodiversity conservation activities [17], an initial review recognised that this community participation was dependent on considerable advice and support [54]. The challenges that communities and individuals face in addressing wicked problems may explain concerns raised in subsequent NZBS reviews that monitoring and reporting of biodiversity conservation activities are patchy or lacking [25,53].

Citizen science volunteer programs are not new to environmental monitoring (e.g., eBird [55]), and are being used increasingly for biodiversity assessment [56]. Citizen science is considered to be a developing field for the collection of long-term field data in ecological restoration [18,56–58]. There are numerous case studies of citizen science activities in "successful" ecological restoration projects (e.g., Tiritiri Matangi Island, New Zealand [59]), indicative of the potential for community-sourced participants to use empirical measures to assess progress towards restoration targets. In addition to the generation of ecological data, citizen science also increases science literacy and offers numerous social benefits as a result of members of the public being engaged collaboratively in research experiences [60–62].

The attributes expected of a restored ecosystem are frequently articulated, but there is little information on whether—or how—these attributes should be measured. The benefits of community participation in biodiversity management are well established (e.g., [4,7,8]), but, as of yet, the relationship between community group aspirations (their goals/aims) and their conservation achievements is unclear. Here, we review the extent to which community groups involved with selected ecological restoration projects in New Zealand define progress for their projects. Specifically, we ask:

1   What elements are covered in the goals and/or aims of the community groups?
2   What activities and strategic milestones do the groups identify?

Finally, we consider whether there are appropriate institutional frameworks in support of their endeavours.

2. Materials and Methods

The study focused on aims and goals of community groups participating in ecological restoration projects, and considered two key questions:

1   What are the key aspirations identified by community groups in their ecological restoration activities?

2   What progress are community groups making towards achieving strategic ecological restoration milestones?

Data about ecological restoration projects were gathered through online internet research. Internet-mediated research (IMR) [63] has the advantage of facilitating fast and efficient access to a broad selection of specialist information [64]. We reviewed ecological restoration projects with community participation through access to targeted websites. We consider that the disadvantages that may be attributed to IMR, (e.g., non-representative samples, ethical issues, and uncertain reliability [64–66]) do not apply to this study, as only text-based data about the targeted organisations was collected, not personal information. Furthermore, the information we accessed is publicly available. In New Zealand, participants in community conservation initiatives often form non-governmental collectives, such as incorporated societies or charitable trusts dedicated to specific conservation projects. This accords legal status to the groups, with an obligation to provide open access to their aims through the New Zealand Companies Office [67]. Groups are required to update their documents annually, so it is assumed that the available information is current and reliable.

The ecological restoration projects investigated covered a wide range of restoration types initiated and actioned by stakeholders at agency, community, and private levels. However, all involved a participatory community group. Seventy-eight New Zealand ecological restoration projects are listed or described by Sanctuaries of New Zealand Inc. [68]. Additional community-based projects were identified through a recent publication [23], the New Zealand Landcare Trust [69], and the authors' knowledge of restoration projects.

Fifty of these projects were reviewed in this study (Figure 1). Forty-one were selected for accessibility of project information through the umbrella websites Sanctuaries of New Zealand [68] and New Zealand Landcare Trust [69]. The websites of other known restoration projects (nine), not listed with the umbrella websites, were targeted individually. A further selection criterion was to limit the geographic range of the projects to the North Island of New Zealand where most of New Zealand's population, and hence the community-based restoration projects, are concentrated [70]. These locations include DOC reserves and other Crown land (71%), island (22%), and mainland sites. The abundance of island projects reflects concentrations of islands around northern New Zealand and their importance as refugia for native species [23,71,72].

The community groups participating in the projects included:

- Volunteer collectives with legal identity, with the project invariably located on public land (these examples include both agency- and community-initiated projects);
- Whānau or hapū (family or sub-tribe of Māori, the indigenous peoples of New Zealand [73]), with the project undertaken on family or tribal land (private);
- Private non-Māori individual(s) undertaking ecological restoration on their own land.

To establish the key elements of the community groups' aspirations, a corpus of comparable data was assembled from their collective goals and aims. This corpus was analysed using Wmatrix [74], a computer-based tool to calculate key-word frequencies. A word cloud visualisation of the dominant words in the corpus was generated using a web-based text analysis tool, Voyant Tools [75]. Word clouds show the frequencies of different words in the corpus as different font sizes, with words at higher frequencies being larger relative to other words. Function words that contribute to sentence syntax rather than meaning were excluded from the analysis. World clouds provide a quick visualisation of the common themes in texts, and are recognised as a supplementary tool for text analyses, particularly for corpora prior to any content manipulation [76].

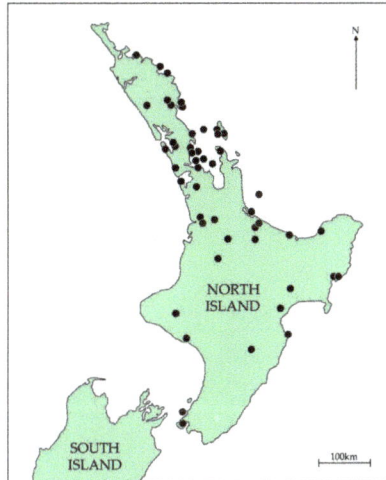

Figure 1. Ecological restoration projects with community participation considered in this study (North Island, New Zealand).

The top key words and multi-word expressions were also analysed for their association with attributes of restored ecosystems. Restoration ecologists have identified the expected outcomes for restoration projects, and empirical measures for the measurement of the progress (or "success") of a restoration process. These broad attributes (or criteria), derived from concepts of ecological integrity, conservation biology, and sustainability, were collated into nine broad groupings (Table 1).

Table 1. Broad attributes of ecological restoration success or progress.

Potential Ecological Restoration Attributes	Literature Sources
Ecosystem representativeness in comparison to reference ecosystem	[77–79]
Ecosystem composition in comparison to reference ecosystem	[77–85]
Maintenance of ecosystem processes	[77–86]
Integration into a larger ecological matrix	[77,79–82,85,86]
Prevention of extinctions and declines	[77–83,85]
Reduction of spread and dominance of alien species	[77–83,85]
Re-establishment of landforms and hydrology	[77,80,82,85]
Educational, scientific, social benefits	[77,82,83,85]
Sustainable management and use	[77,81,82]

Based on the information reported through the projects' websites, an evaluation of the progress made towards strategic ecological restoration milestones was carried out by scoring each project on an ordinal scale to characterise the status (Table 2). This evaluation included noting the nature of any ecological monitoring activities that were being reported.

Table 2. Scoring criteria for characterisation of the monitoring status of ecological restoration projects.

Project Status	Score				
	1	2	3	4	5
No project aims identifiable	✔				
Project aims identifiable		✔	✔	✔	✔
Evidence of monitoring of populations of invasive species			✔	✔	✔
Evidence of monitoring of populations of indigenous species				✔	✔
Evidence of monitoring of populations of indigenous species; evidence of ecological monitoring to show restoration progress					✔

3. Results

3.1. Projects' Aims and Objectives

The projects' objectives, where available, were collated into a corpus of 1490 words and are presented in visual form in Figure 2.

Figure 2. Visualisation of the dominant words in the goals and objectives of 50 community groups participating in ecological restoration projects. Word size reflects frequency of occurrence.

The key words in the corpus were: "native" (1.61%), "island" (0.87%), "natural" (0.87%), "restore" (0.87%), "provide" (0.81%), "conservation" (0.74%), "ecosystem" (0.60%), "education" (0.60%), "flora" (0.60%), "fauna" (0.60%), "species" (0.60%), and "birds" (0.54%). The two top multi-word expressions in the corpus, "endangered species" and "pest control", were equal in their occurrence (0.13%).

Words associated with the formal disciplines of conservation biology and restoration ecology were present, but at much lower frequencies: e.g., "indigenous" (0.27%), "habitat" (0.27%), "monitoring"(0.2%), "ecosystems" (0.2%), "research" (0.2%), "scientific" (0.13%), "populations" (0.13%), "reintroduction" (0.13%), "ecological" (0.13%), and "ecology" (0.13%).

The alignment of the key words and multi-word expressions with broad attributes of ecological restoration are shown in Table 3.

Table 3. Top 10 key words and top 2 multi-word expressions from community groups' aims and objectives aligned with attributes of restored ecosystems.

Potential Ecological Restoration Attributes	Literature Sources	Aims and Objective Key Words/Phrases
Ecosystem representativeness in comparison to reference ecosystem	[77–79]	native, natural, island, restore, ecosystem, flora, fauna, species, birds
Ecosystem composition in comparison to reference ecosystem	[77–85]	native, natural, island, restore, ecosystem, education, flora, fauna, species, birds
Maintenance of ecosystem processes	[77–86]	natural, provide, ecosystem, flora, fauna, species, birds
Integration into a larger ecological matrix	[77,79–82,85,86]	natural, island, restore, provide, ecosystem, species
Prevention of extinctions and declines	[77–83,85]	native, island, restore, conservation, flora, fauna, species, birds, endangered species
Reduction of spread and dominance of alien species	[77–83,85]	native, restore, provide, conservation, ecosystem, education, flora, fauna, species, pest control
Re-establishment of landforms and hydrology	[77,80,82,85]	restore, ecosystem
Educational, scientific, social benefits	[77,78,82,83,85]	conservation, education
Sustainable management and use	[77,78,82,83]	conservation, ecosystem, education

3.2. Project Milestones

The evaluation of the progress made towards strategic ecological restoration milestones, based on the information provided through the projects' websites, is illustrated in Figure 3. The project status is based on the scoring system described in Table 2. No groups demonstrated evidence of comprehensive ecological monitoring to indicate progress towards a pre-determined restoration state.

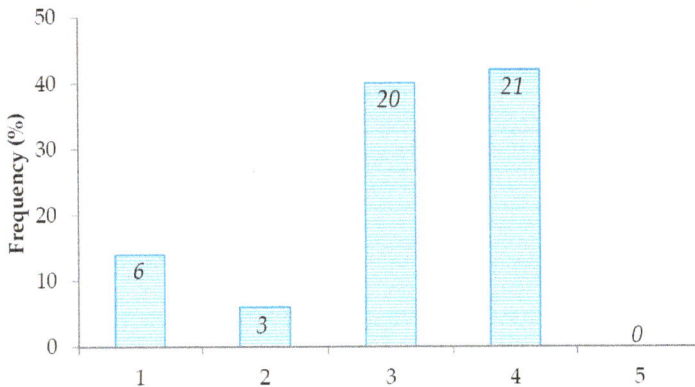

Figure 3. Proportion of projects achieving ecological restoration milestones (n = 50; italicised numerals within bars indicate the number of groups at each status).

The reporting of monitoring activities indicated that a majority of the groups (43) were engaged in control or monitoring of introduced mammals. Species reintroduction as a past, or intended, activity was identified by 31 of the groups. Of the groups that had actually completed species translocations (25), only 13 indicated that monitoring of the translocated species was being undertaken.

4. Discussion

4.1. Elements Covered in the Goals and/or Aims of the Community Groups

The goals and aims of the ecological restoration groups in the study varied greatly in the quality of information provided. The web-based information contained shortcomings in the use of restoration ecology terminology, and no groups identified specific strategic milestones for their participation and progress towards them. Our study confirms, however, that numerous community groups are involved in ecological restoration activities that fit the concept of "wicked" problems. Most groups are engaged in the management of mammal pests (84%), with species reintroduction either being carried out, or on the "wish list", for 62% of the groups. Both of these activities are complex, may have uncertain outcomes, and are potentially controversial.

The key words that emerged from the analysis of the aims and objectives of 50 community collectives reinforce studies identifying environmental gain as a significant motivational driver of volunteers participating in ecological restoration projects [81]. These key words and phrases were, however, of a relatively general nature, often with a local focus, and did not use the formal language more often associated with the science of restoration ecology. Examples of such generalised aims or objectives are:

- "to restore the natural and cultural landscapes" [87];
- "future generations will enjoy a forest alive with native birds, reptiles and insects" [88];
- "a natural environment of indigenous flora and fauna" [89];
- "a corridor of bush along a pristine stream; a place for birds to live and kids to play [90];
- "to preserve and enhance the natural beauty, ecosystems and biodiversity" [91];
- "to remove forever, introduced mammalian pests and predators" [92].

The generalised aims and objectives may reflect a gap identified in other studies (e.g., [52]) between a professional perspective of ecological restoration (e.g., that of a government agency, with a legislative obligation to focus on the scientific foundations of restoration), and a community perspective where general restoration objectives support social environmental benefits that avocational participants relate to. Avocational volunteers (or lay ecological restorationists [16,93]) have been shown to conceptualise restoration differently to ecological professionals [52]. However, we believe that, despite the general language used in the community groups' goals and aims, the key words and multi-word expressions can still be aligned to the broad attributes of ecological restoration (Table 3).

Words associated with long-term monitoring of ecological attributes, however, were few, and, if present, were at particularly low frequency. Although monitoring featured in a minority of objectives, words that might be expected of community-based restoration projects, such as "citizen", "science", and "measure", were absent. This low focus on ecological science likely reflects the avocational status of the community participants who generally lack technical skills and science literacy, particularly where their contribution focuses on practical contribution under the direction of an agency body [94].

4.2. Activities and Strategic Milestones Identified by Community Groups

Environmental monitoring is a globally common participatory action (e.g., [4,7,8,94]). However, although it is evident that some degree of monitoring is undertaken by the community groups we studied, none provide a comprehensive framework of strategic milestones to demonstrate restoration progress. Other studies [94] have found similarly that only a minority of groups monitor the outcome of their actions.

Analyses of reports from restoration projects that we studied indicated long term monitoring activities dominated by assessment of ecological status (e.g., populations) and impacts (e.g., invasive mammals). Aspects of ecological integrity, such as ecosystem composition, structure, and processes, do not feature in the analyses. Ecological status assessment is dominated by vegetation measures, attributed to the relative speed and ease with which such measures can be completed [95,96]. Recent publications suggest that this is a common characteristic of restoration monitoring [94,96]. The milestones that were used by the groups in our study focussed on the monitoring of invasive mammal pests (see also [24]). This form of monitoring reflects the impact of invasive mammals in New Zealand, and that their management is now a routine activity and acknowledged as a crucial first phase (and potentially long-term activity) of restoration projects [97]. Of the more specific restoration activity of species translocation, only half of the groups that had carried out, or participated in, such events indicated that these species were monitored (despite this being a condition of all translocation permits issued by the Department of Conservation).

There is evidence that many ecological restoration projects do exhibit positive progress in improving ecological integrity and/or achieving conservation gains [98], and this appears to be generally assumed for a large number of New Zealand projects [23,99]. Despite such proclamations of successful ecological restoration projects worldwide, however, there is international debate over what constitutes "success" in the context of ecological restoration [98,100–104], with some authors criticising the wide use of the word "success" (and other value-laden terms, for example, "desirable", "degraded", and "intact") to describe ecological restoration outcomes (e.g., [98,100,103]). An additional complication, further reducing clarity for community participation in particular, is that different stakeholders will likely have differing expectations of restoration outcomes [105], so it is inevitable that perceptions of restoration success will be based on their respective experience and values [98,100,105].

If the measurement of restoration progress and/or success is a desired and expected component of restoration outcomes, then the need to assign value to the actions of community participation is paramount. Accountability is particularly important where restoration projects are on public lands and restoration outcomes are accountable to society (taxpayers) in general, and/or where parties using outside funding may be held financially responsible for restoring a damaged ecosystem [106]. Effective and meaningful evaluation of the progress of an ecological restoration project requires a framework for systematic ecological monitoring to be in place [83]. Such a framework is deemed to be a necessary pre-requisite for evidence-based review of restoration actions [107].

4.3. Citizen Science

New Zealand has international recognition for the participation of volunteers in conservation projects [38,59], with the growing trend of volunteers seen by the Department of Conservation and other territorial authorities as a pathway to greater engagement of the public in conservation activities and to increase business partnerships for conservation gain (DOC 2013). This trend addresses the goals of the NZBS [17], and is consistent with current government priorities [16]. Since communities may already be engaged in some level of monitoring, albeit at a low scale, the targeted training of interested individuals as citizen scientists may prove a way to obtain comprehensive measures of restoration outcomes against community effort. However, citizen science is not listed anywhere in the NZBS [17], nor in the index of a comprehensive published account of New Zealand sanctuaries [23]. Nevertheless, the Strategy emphasises the need to improve the technical knowledge and capacity for communities to become involved in biodiversity management.

Citizen science, however, has its problems. It requires willing engagement of participants in science. This is a likely barrier, as, according to published accounts (e.g., [10,108]), most community volunteers are involved in the projects for reasons other than an interest in obtaining data. In addition, volunteers' experiences must be enjoyable to be sustainable [21], and the imposition of science-based activities may counter this need, particularly for avocational participants. The divide between avocational participants and formal science is well recognised [109], and may have developed through the marginalisation of amateur naturalists as a by-product of the professionalisation of science [18]. Furthermore, the quality of data collected by avocational volunteers has often been questioned [58], although recent approaches tend to suggest that citizen science data is useful and important if the research methodologies are well-designed [60,110].

The involvement of avocational volunteers in the measurement of ecological attributes of restoration, measures that are essentially an applied science focus, has implications for both achieving and measuring project outcomes. For many participants in restoration projects, social and recreational motivations may be as important as environmental stewardship [111], and are often included in the goals of citizen science projects. Their participation in ecological science, therefore, may be unintentional, and develop as a result of a project's management requirements and/or devolution of management by governing agencies to communities. In this situation, the participants may be considered as "accidental" scientists rather than citizen scientists [112], and raises questions of the desire, and hence capacity, of community groups to engage in ecological science.

5. Conclusions

This study considered 50 community-based ecological restoration projects in New Zealand to gauge the relationship between the aspirations of the participants as collectives (their goals/aims) and their conservation achievements. We found that goals tend to be generalised, and do not identify strategic milestones to gauge project success. Although many groups are undertaking environmental monitoring and, at least at some level, are engaged effectively in scientific activities, it appears that monitoring of restoration outcomes is given a lower priority than might be expected given the need to provide measures of long-term results, which perhaps is indicative of the avocational nature of most participants.

The consequence of poorly defined criteria of restoration "success" is the difficulty of evaluating benefits, leading to subjective assertions of the outcomes of ecological restoration, particularly as they are often based on anecdotal measures. This difficulty can be compounded when projects lack predetermined goals, criteria for measuring milestones, and fail to monitor appropriate project outcomes [98]. Evidence-based approaches would justify restoration as an option for natural resource management through demonstrated conservation gains, development of best practices in the field, facilitated prioritisation of restoration actions, justified funding and allocation of resources, and clear accounting for funds committed to do the work [98,104,113–115].

Citizen science is accepted as an excellent opportunity to progress ecological restoration, where the participants increase their scientific literacy and skills, and gain social benefit. Such activities can certainly meet the needs of effective monitoring for restoration success. However, although most studies of community groups identify the need for increased technical training for volunteers, including methods for outcome monitoring, such training is not necessarily a priority of the groups themselves [24,116].

The implication from the NZBS [17] is that improved capability through training and technical support would derive from professional agencies such as the New Zealand Department of Conservation (DOC) and regional councils. Many community groups do receive assistance and advice from staff of government agencies or tertiary educational institutes [24], sometimes with scientists as members or even instigators of the community groups. Such collaborative processes have been shown to greatly enhance the outcomes for conservation [117]. However, studies of community participation in ecological restoration do not mention the collaborative approach as a motivation for involvement, although it is perhaps buried within the objective of education and learning about the environment [24]. Furthermore, two events have conspired to make collaboration increasingly difficult to achieve. First, the number of community groups is now so large it is potentially beyond the agencies' abilities to help them all. Second, technical assistance from DOC has declined as funding has progressively reduced [23]. In an attempt to improve linkages with the community, DOC has bolstered partnership staff, but in compensation further reduced its technical capacity. These two events are mirrored globally (e.g., [10,62,118,119]).

Ecological restoration involving community groups thus faces a conundrum: participation is an essential component of the goals of conservation agencies. Restoration participants, however, may engage in citizen science if they are already interested in science. If not, many participants, at best, are "accidental" scientists rather than citizen scientists. There are important implications for managers of existing and future citizen-science projects. Given the "wicked" problems associated with ecological restoration activities—technical, ethical, and financial—community groups will need more than intermittent technical advice from government agencies. These groups will require substantial training in data gathering and analyses for citizen scientists, with support from institutions and innovative tools, in order to generate the long-term resilience necessary for sustainable ecological restoration projects.

Acknowledgments: We extend our appreciation to John Perrott (Auckland University of Technology) for the discussions surrounding the concept of accidental scientists. We thank Rebecca Jarvis, Graham Jones, Angela Dale, and the three anonymous reviewers for their constructive comments of the draft manuscript. This study and its publication costs were covered through Auckland University of Technology Faculty of Health and Environmental Sciences PhD Scholarship funding, and research support from Unitec Institute of Technology.

Author Contributions: David Towns and Barbara Breen conceived the project; Mel Galbraith carried out the review and wrote the paper.

Conflicts of Interest: The authors declare no conflict of interest. The funding sponsors had no role in the design of the study; in the collection, analyses, or interpretation of data; in the writing of the manuscript, nor in the decision to publish the results.

Abbreviations

The following abbreviations are used in this manuscript:

DOC   New Zealand Department of Conservation
IMR   Internet-Mediated Research
NGO   Non-Governmental Organisation
NZ   New Zealand
NZBS   New Zealand Biodiversity Strategy

References

1. Quick, K.S.; Bryson, J.M. Public participation. In *Handbook in Theories of Governance*; Torbing, J., Ansell, C., Eds.; Edward Elgar Press: Cheltenham, UK, 2016; pp. 158–169.
2. Barton, B. Underlying concepts and theoretical issues in public participation in resources development. In *Human Rights in Natural Resource Development: Public Participation in the Sustainable Development of Mining and Energy Resources*; Zillman, D.M., Lucas, A., Pring, G., Eds.; Oxford University Press: New York, NY, USA, 2002; pp. 77–120.
3. Arnstein, S.R. A Ladder of citizen participation. *J. AM. Plan. Assoc.* **1969**, *35*, 216–224. [CrossRef]
4. Forgie, V.; Horsley, P.; Johnson, J. *Facilitating Community-Based Conservation Initiatives*; Department of Conservation: Wellington, New Zealand, 2001.
5. Creighton, J.L. *The Public Participation Handbook: Making Better Decisions Through Citizen Involvement*; John Wiley & Sons: Chichester, UK, 2005.
6. Claridge, T. Social Capital and Natural Resource Managment. Master's Thesis, University of Queensland, Brisbane, Australia, July 2004.
7. Bixler, R.P.; Dell'Angelo, J.; Mfune, O.; Roba, H. The political ecology of participatory conservation: Institutions and discourse. *J. Political Ecol.* **2015**, *22*, 164–182.
8. Ockenden, N. *Volunteering in the Natural Outdoors in the UK and Ireland: A Literature Review*; Institute for Volunteering Research: London, UK, 2007.
9. Evely, A.C.; Pinard, M.; Reed, M.S.; Fazey, I. High levels of participation in conservation projects enhance learning. *Conserv. Lett.* **2011**, *4*, 116–126. [CrossRef]
10. Bramston, P.; Pretty, G.; Zammit, C. Assessing environmental stewardship motivation. *Environ. Behav.* **2011**, *43*, 776–788. [CrossRef]
11. Miles, I.; Sullivan, W.C.; Kuo, F.C. Ecological restoration volunteers: The benefits of participation. *Urban Ecosyst.* **1998**, *2*, 27–41. [CrossRef]
12. Cheng, A.S.; Sturtevant, V.E. A framework for assessing collaborative capacity in community-based public forest management. *Environ. Manag.* **2011**, *49*, 675–689. [CrossRef] [PubMed]
13. Grese, R.E.; Kaplan, R.S.; Ryan, R.L.; Buxton, J. Psychological benefits of volunteering in stewardship programmes. In *Restoring Nature: Perspectives from the Social Sciences and Humanities*; Gobster, P.H., Hull, R.B., Eds.; Island Press: Washington, DC, USA, 2001; pp. 265–280.
14. Hardie-Boys, N. *Valuing Community Group Contributions to Conservation*; Department of Conservation: Wellington, New Zealand, 2010.
15. Miles, I.; Sullivan, W.C.; Kuo, F.E. Psychological benefits of volunteering for restoration projects. *Ecol. Restor.* **2000**, *18*, 218–227.

16. State Services Commission. *Performance Improvement Framework: Review of the Department of Conservation (DOC)*; New Zealand Government: Wellington, New Zealand, 2014.

17. Anon. *New Zealand Biodiversity Strategy: Our Chance to Turn the Tide*; Department of Conservation and Ministry for the Environment: Wellington, New Zealand, 2000.

18. Miller-Rushing, A.; Primack, R.; Bonney, R. The history of public participation in ecological research. *Front. Ecol. Environ.* **2012**, *10*, 285–290. [CrossRef]

19. Cornwell, M.L.; Campbell, L.M. Co-producing conservation and knowledge: Citizen-based sea turtle monitoring in North Carolina, USA. *Soci. Stud. Sci.* **2012**, *42*, 101–120. [CrossRef]

20. Ellis, R.; Waterton, C. Caught between the cartographic and the ethnographic imagination: The whereabouts of amateurs, professionals, and nature in knowing biodiversity. *Environ. Plan. Soc. Sp.* **2005**, *23*, 673–693. [CrossRef]

21. Bell, K. *Assessing the Benefits for Conservation of Volunteer Involvement in Conservation Activities*; Department of Conservation: Wellington, New Zealand, 2003.

22. New Zealand Plant Conservation Network. Available online: http://www.nzpcn.org.nz/page.aspx?conservation_restoration_find_a_group (accessed on 20 Janurary 2016).

23. Butler, D.; Lindsay, T.; Hunt, J. *Paradise Saved: The Remarkable Story of New Zealand's Wildlife Sanctuaries and How They are Stemming the Tide of Extinction*; Random House: Auckland, New Zealand, 2014.

24. Peters, M.A.; Hamilton, D.; Eames, C. Action on the ground: A review of community environmental groups' restoration objectives, activities and partnerships in New Zealand. *NZ J. Ecol.* **2015**, *39*, 179–189.

25. Green, W.; Clarkson, B.D. *Turning the Tide? A Review of the First Five Years of the New Zealand Biodiversity Strategy: The Synthesis Report*; Department of Conservation: Wellington, New Zealand, 2005.

26. Berkes, F. Rethinking community-based conservation. *Conserv. Biol.* **2004**, *18*, 621–630. [CrossRef]

27. Towns, D.R. Eradications of vertebrate pests from islands around New Zealand: What have we delivered and what have we learned? In *Island Invasives: Eradication and Management*; Veitch, C.R., Clout, M.N., Towns, D.R., Eds.; IUCN: Gland, Switzerland, 2011; pp. 364–371.

28. Miskelly, C.M.; Powlesland, R.G. Conservation translocations of New Zealand birds, 1863–2012. *Notornis* **2013**, *60*, 3–28.

29. Seddon, P.J.; Maartin Strauss, W.; Innes, J. Animal translocations: What are they and why do we do them? In *Reintroduction Biology: Integrating Science and Management*; Ewen, J.G., Armstrong, D.P., Parker, K.A., Seddon, P.J., Eds.; John Wiley & Sons: Chichester, UK, 2012; pp. 1–32.

30. Kowarik, I. Time lags in biological invasions with regard to the success and failure of alien species. In *Plant Invasions—General Aspects and Special Problems*; Pyšek, P., Prach, K., Rejmánek, M., Wade, M., Eds.; SPB Academic Publishing: Amsterdam, The Netherland, 1995; pp. 15–39.

31. Parkes, J.P.; Panetta, F.D. Eradication of invasive species: Progress and emerging issues in the 21st century. In *Invasive Species Management: A Handbook of Principles and Techniques*; Clout, M.N., Williams, P.A., Eds.; Oxford University Press: Oxford, UK, 2009.

32. Furness, R.W.; Monaghan, P. *Seabird Ecology*; Blackie & Son: Glasgow, UK, 1987.

33. Towns, D.R.; Byrd, G.V.; Jones, H.P.; Rauzon, M.J.; Russell, J.C.; Wilcox, C. Impacts of introduced predators on seabirds. In *Seabird Islands: Ecology, Invasions, and Restoration*; Mulder, C.P.H., Anderson, W.B., Towns, D.R., Bellingham, P.J., Eds.; Oxford University Press: Oxford, UK, 2011; pp. 56–90.

34. Grice, T. Principles of containment and control of invasive species. In *Invasive Species Management: A Handbook of Principles and Techniques*; Clout, M.N., Williams, P.A., Eds.; Oxford University Press: Oxford, UK, 2009; pp. 61–76.

35. Russell, J.C.; Towns, D.R.; Clout, M.N. *Review of Rat Invasion Biology: Implications for Island Biosecurity*; Department of Conservation: Wellington, New Zealand, 2008.

36. Larson, D.L.; Phillips-Mao, L.; Quiram, G.; Sharpe, L.; Stark, R.; Sugita, S.; Weiler, A. A framework for sustainable invasive species management: Environmental, social, and economic objectives. *J. Environ. Manag.* **2011**, *92*, 14–22. [CrossRef] [PubMed]

37. Towns, D.R.; West, C.J.; Broome, K.G. Purposes, outcomes and challenges of eradicating invasive mammals from New Zealand islands: An historical perspective. *Wildl. Res.* **2013**, *40*, 94–107. [CrossRef]

38. Towns, D.R.; Aguirre-Muñoz, A.; Kress, S.W.; Hodum, P.J.; Burbidge, A.A.; Saunders, A. The social dimension—Public involvement in seabird island restoration. In *Seabird Islands: Ecology, Invasion, and Restoration*; Mulder, C.P.H., Anderson, W.B., Towns, D.R., Bellingham, P.J., Eds.; Oxford University Press: Oxford, UK, 2011; pp. 358–392.

39. Russell, J.C.; Broome, K.G. Fifty years of rodent eradications in New Zealand: Another decade of advances. *NZ J. Ecol.* **2016**, *40*, 197–204. [CrossRef]

40. Armstrong, D.P.; Hayward, M.W.; Moro, D.; Seddon, P.J. Introduction: The development of reintroduction biology in New Zealand and Australia. In *Advances in Reintroduction Biology of Australian and New Zealand Fauna*; Armstrong, D.P., Hayward, M.W., Moro, D., Seddon, P.J., Eds.; CSIRO Publishing: Melbourne, Australia, 2015.

41. Ewen, J.G.; Armstrong, D.P.; Parker, K.A.; Seddon, P.J. *Reintroduction Biology: Integrating Science and Management*; John Wiley & Sons: Chichester, UK, 2012.

42. Seddon, P.J.; Armstrong, D.P.; Parker, K.A.; Ewen, J.G. Summary. In *Reintroduction Biology: Integrating Science and Management*; Ewen, J.G., Armstrong, D.P., Parker, K.A., Seddon, P.J., Eds.; John Wiley & Sons: Chichester, UK, 2012; pp. 476–481.

43. Jamieson, I.G.; Lacy, R.C. Managing genetic issues in reintroduction biology. In *Reintroduction Biology: Integrating Science and Management*; Ewen, J.G., Armstrong, D.P., Parker, K.A., Seddon, P.J., Eds.; John Wiley & Sons: Chichester, UK, 2012; pp. 441–475.

44. Nichols, J.D.; Armstrong, D.P. Monitoring for reintroductions. In *Reintroduction Biology: Integrating Science and Management*; Ewen, J.G., Armstrong, D.P., Parker, K.A., Seddon, P.J., Eds.; John Wiley & Sons: Chichester, UK, 2012; pp. 223–255.

45. Parker, K.A.; Ewen, J.G.; Seddon, P.J.; Armstrong, D.P. Post-release monitoring of bird translocations: Why is it important and how do we do it? *Notornis* **2013**, *60*, 85–92.

46. Cromarty, P.L.; Alderson, S.L. Translocation statistics (2002–2010), and the revised department of conservation translocation process. *Notornis* **2013**, *60*, 55–62.

47. Nally, S.; Adams, L. Evolution of the translocation approval process in Australia and New Zealand. In *Advances in Reintroduction Biology of Australian and New Zealand Fauna*; Armstrong, D.P., Hayward, M.W., Moro, D., Seddon, P.J., Eds.; CSIRO Publishing: Melbourne, Australia, 2015; pp. 273–284.

48. Parker, K.A. Translocations: Providing outcomes for wildlife, resource managers, scientists, and the human community. *Restor. Ecol.* **2008**, *16*, 204–209. [CrossRef]

49. Rittel, H.W.J.; Webber, M.M. Dilemmas in a general theory of planning. *Policy Sci.* **1973**, *4*, 155–169. [CrossRef]

50. Ludwig, D. The era of Management is over. *Ecosystems* **2001**, *4*, 758–764. [CrossRef]

51. Reed, M.S. Stakeholder participation for environmental management: A literature review. *Biol. Conser.* **2008**, *141*, 2417–2431. [CrossRef]

52. Weng, Y. Contrasting visions of science in ecological restoration: Expert-Lay dynamics between professional practitioners and volunteers. *Geoforum* **2015**, *65*, 134–145. [CrossRef]

53. Green, W.; Clarkson, B.D. *Review of the New Zealand Biodiversity Strategy Themes*; Department of Conservation: Wellington, New Zealand, 2006.

54. Anon. *New Zealand Biodiversity Strategy: Third Annual Report 2002/03*; Department of Conservation: Wellington, New Zealand, 2003.

55. Sullivan, B.L.; Wood, C.L.; Iliff, M.J.; Bonney, R.E.; Fink, D.; Kelling, S. eBird: A citizen-based bird observation network in the biological sciences. *Biol. Conser.* **2009**, *142*, 2282–2292. [CrossRef]

56. Couvet, D.; Prevot, A. Citizen-science programs: Towards transformative biodiversity governance. *Environ. Dev.* **2015**, *13*, 39–45. [CrossRef]

57. Bonney, R.; Cooper, C.B.; Dickinson, J.; Kelling, S.; Phillips, T.; Rosenberg, K.V.; Shirk, J.L. Citizen science: A developing tool for expanding science knowledge and scientific literacy. *BioScience* **2009**, *59*, 977–984. [CrossRef]

58. Bonney, R.; Shirk, J.L.; Phillips, T.B.; Wiggins, A.; Ballard, H.L.; Miller-Rushing, A.J.; Parrish, J.K. Next steps for citizen science. *Science* **2014**, *343*, 1436–1437. [CrossRef] [PubMed]

59. Galbraith, M. Public and ecology—The role of volunteers on Tiritiri Matangi. *NZ J. Ecol.* **2013**, *37*, 266–271.

60. Dickinson, J.L.; Shirk, J.L.; Bonter, D.; Bonney, R.; Crain, R.L.; Martin, J.; Phillips, T.; Purcell, K. The current state of citizen science as a tool for ecological research and public engagement. *Front. Ecol. Environ.* **2012**, *10*, 291–297. [CrossRef]

61. Dickinson, J.L.; Bonney, R. *Introduction: Why citizen science? In Citizen Science: Public Participation in Environmental Research*; Dickinson, J.L., Bonney, R., Eds.; Cornell University Press: Ithaca, NY, USA, 2012; pp. 1–14.

62. Conrad, C.C.; Hilchey, K.G. A review of citizen science and community-based environmental monitoring: Issues and opportunities. *Environ. Monit. Assess.* **2011**, *176*, 273–291. [CrossRef] [PubMed]

63. Hewson, C.; Yule, P.; Laurent, D.; Vogel, C. *Internet Research Methods: A Practical Guide for the Social and Behavioural Sciences*; SAGE Publications: London, UK, 2003.

64. Hewson, C.; Vogel, C.; Laurent, D. *Internet Research Methods*; Sage Publishing: London, UK, 2016.

65. Hewson, C.; Laurent, D. Research design and tools for internet research. In *Sage Handbook of Online Research Methods*; Lee, R.M., Fielding, N., Blank, G., Eds.; SAGE Publications: London, UK, 2008; pp. 58–78.

66. Fletcher, W.H. Corpus analysis of the world wide web. In *The Encyclopedia of Applied Linguistics*; Chapelle, C., Ed.; John Wiley & Sons: Chichester, UK, 2012.

67. Companies Office. What are the Advantages of Becoming an Incorporated Society? Available online: http://www.societies.govt.nz/cms/customer-support/faqs/incorporated-societies/what-are-the-advantages-of-becoming-an-incorporated-society (accessed on 6 February 2016).

68. Sanctuaries of New Zealand Inc. Available online: http://www.sanctuariesnz.org/projects.asp (accessed on 13 May 2016).

69. New Zealand Landcare Trust. Available online: http://www.landcare.org.nz/ (accessed on 13 May 2016).

70. Statistics New Zealand. Subnational Population Estimates: At 30 June 2015 (provisional). Available online: http://www.stats.govt.nz (accessed on 21 June 2016).

71. Towns, D.R.; Atkinson, I.A.E.; Daugherty, C.H. Ecological restoration of New Zealand islands—Introduction. In *Ecological Restoration of New Zealand Islands*; Towns, D.R., Daugherty, C.H., Atkinson, I.A.E., Eds.; Department of Conservation: Wellington, New Zealand, 1990; pp. III–IV.

72. Wright, A.E.; Beever, R.E. Introduction. In *The Offshore Islands of New Zealand*; Wright, A.E., Beever, R.E., Eds.; Department of Lands and Survey: Wellington, New Zealand, 1986.

73. Taonui, R. Tribal organisation—The Significance of iwi and hapū. Available online: http://www.TeAra.govt.nz/en/tribal-organisation/page-1 (accessed on 17 May 2016).

74. Rayson, P. Wmatrix: A web-based corpus processing environment. Available online: http://ucrel.lancs.ac.uk/wmatrix/ (accessed on 21 June 2016).

75. Sinclair, S.; Rockwell, G. Voyant Tools. Available online: http://docs.voyant-tools.org/ (accessed on 3 May 2016).

76. McNaught, C.; Lam, P. Using wordle as a supplementary research tool. *Qual. Rep.* **2010**, *15*, 630–643.

77. Lee, W.G.; McGlone, M.; Wright, E.F. *Biodiversity Inventory and Monitoring a Review of National and International Systems and a Proposed Framework for Future Biodiversity Monitoring by the Department of Conservation*; Landcare Research: Wellington, New Zealand, 2005.

78. Lee, W.G.; Allen, R.B. *Recommended Monitoring Framework for Regional Councils Assessing Biodiversity Outcomes in Terrestrial Ecosystems*; Landcare Research: Dunedin, New Zealand, 2011.

79. Towns, D.R.; Wright, E.F.; Stephens, T. Systematic measurement of effectiveness for conservation of biodiversity on New Zealand islands. In Proceedings of the Conserv-Vision conference, Hamilton, New Zealand, 2–4 July 2007.

80. Society for Ecological Restoration Australasia. National Standards for the Practice of Ecological Restoration in Australia. Available online: http://www.seraustralasia.com/standards/contents.html (accessed on 4 May 2016).

81. Society for Ecological Restoration International Science and Policy Working Group. In *The SER Primer on International Ecological Restoration*; Society for Ecological Restoration International: Tucson, AZ, USA, 2004.

82. Keenleyside, K.A.; Dudley, N.; Cairns, S.; Hall, C.M.; Stolton, S. *Ecological Restoration for Protected Areas: Principles, Guidelines and Best Practices*; IUCN: Gland, Switzerland, 2012.

83. Atkinson, I.A.E. *Guidelines to the Development and Monitoring of Ecological Restoration Programmes*; Department of Conservation: Wellington, New Zealand, 1994.

84. Chaves, R.B.; Durigan, G.; Brancalion, P.H.S.; Aronson, J. On the need of legal frameworks for assessing restoration projects success: New perspectives from São Paulo state (Brazil). *Restor. Ecol.* **2015**, *23*, 754–759. [CrossRef]

85. Parks Canada; Canadian Parks Council. *Principles and Guidelines for Ecological Restoration in Canada's Protected Natural Areas*; National Parks Directorate, Parks Canada Agency: Gatineau, QB, Canada, 2008.

86. Ruiz-Jaén, M.C.; Aide, T.M. Vegetation structure, species diversity, and ecosystem processes as measures of restoration success. *For. Ecol. Manag.* **2005**, *218*, 159–173. [CrossRef]

87. Motutapu Restoration Trust. Available online: http://www.motutapu.org.nz/ (accessed on 3 May 2016).

88. Motuora Restoration Society. Available online: http://motuora.org.nz/ (accessed on 3 May 2016).

89. Motuihe Trust. Available online: http://www.motuihe.org.nz (accessed on 3 May 2016).

90. Waitao-Kaiate Environmental Group. Available online: http://www.landcare.org.nz/Landcare-Community/Waitao-Kaiate-Environmental-Group (accessed on 3 May 2016).

91. Whakaangi Landcare Trust. Available online: http://whakaangi.kiwi/ (accessed on 3 May 2016).

92. Maungatautari Ecological Island Trust. Available online: http://www.sanctuarymountain.co.nz/ (accessed on 3 May 2016).

93. New Zealand National Party. *Policy 2014: Conservation*; New Zealand National Party: Wellington, New Zealand, 2014.

94. Peters, M.A.; Hamilton, D.; Eames, C.; Innes, J.; Mason, N.W.H. The current state of community-based environmental monitoring in New Zealand. *NZ J. Ecol.* **2016**, *40*, 279–288. [CrossRef]

95. Young, T.P. Restoration ecology and conservation biology. *Biol. Conser.* **2000**, *92*, 73–83. [CrossRef]

96. Ruiz-Jaén, M.C.; Aide, T.M. Restoration success: How is it being measured? *Restor. Ecol.* **2005**, *13*, 569–577. [CrossRef]

97. Clout, M.N. Biodiversity conservation and the management of invasive animals in New Zealand. In *Invasive Species and Biodiversity Management*; Sandlund, O.T., Schei, P.J., Viken, Å., Eds.; Springer: Dordrecht, The Netherland, 2001; pp. 349–361.

98. Suding, K.N. Toward an era of restoration in ecology: Successes, failures, and opportunities ahead. *Annu. Rev. Ecol. Evol. Syst.* **2011**, *42*, 465–487. [CrossRef]

99. Campbell-Hunt, D.; Campbell-Hunt, C. *Ecosanctuaries: Communities Building a Future for New Zealand's Threatened Ecologies*; Otago University Press: Dunedin, New Zealand, 2013.

100. Clewell, A.; Rieger, J.P. What practitioners need from restoration ecologists. *Restor. Ecol.* **1997**, *5*, 350–354. [CrossRef]

101. Palmer, M.; Bernhardt, E.S.; Allan, J.D.; Lake, P.S.; Alexander, G.; Brooks, S.; Carr, J.; Clayton, S.; Dahm, C.N.; Follstad Shah, J.; et al. Standards for ecologically successful river restoration. *J. Appl. Ecol.* **2005**, *42*, 208–217. [CrossRef]

102. Bernhardt, E.S.; Sudduth, E.B.; Palmer, M.A.; Allan, J.D.; Meyer, J.L.; Alexander, G.; Follastad-Shah, J.; Hassett, B.; Jenkinson, R.; Lave, R.; et al. Restoring rivers one reach at a time: Results from a survey of U.S. river restoration practitioners. *Restor. Ecol.* **2007**, *15*, 482–493. [CrossRef]

103. Zedler, J.B. Success: An unclear, subjective descriptor of restoration outcomes. *Ecol. Restor.* **2007**, *25*, 162–168. [CrossRef]

104. Wortley, L.; Hero, J.-M.; Howes, M. Evaluating ecological restoration success: A review of the literature. *Restor. Ecol.* **2013**, *21*, 537–543. [CrossRef]

105. Hackney, C.T. Restoration of coastal habitats: Expectation and reality. *Ecol. Eng.* **2000**, *15*, 165–170. [CrossRef]

106. Holl, K.D.; Howarth, R.B. Paying for restoration. *Restor. Ecol.* **2000**, *8*, 260–267. [CrossRef]

107. Sutherland, W.J.; Pullin, A.S.; Dolman, P.M.; Knight, T.M. The need for evidence-based conservation. *Trends Ecol. Evol.* **2004**, *19*, 305–308. [CrossRef] [PubMed]

108. Bruyere, B.; Rappe, S. Identifying the motivations of environmental volunteers. *J. Environ. Plan. Manag.* **2007**, *50*, 503–516. [CrossRef]

109. Braunisch, V.; Home, R.; Pellet, J.; Arlettaz, R. Conservation science relevant to action: A research agenda identified and prioritized by practitioners. *Biol. Conserv.* **2012**, *153*, 201–210. [CrossRef]

110. Cohn, J.P. Citizen science: Can volunteers do real research? *BioScience* **2008**, *58*, 192–197. [CrossRef]

111. Reid, K.A.; Williams, K.J.H.; Paine, M.S. Hybrid knowledge: Place, practice, and knowing in a volunteer ecological restoration project. *Ecol. Soc.* **2011**, *16*, 19. [CrossRef]

112. Perrott, J.K. Auckland University of Technology, Auckland, New Zealand. Personal communication, 2015.

113. Palmer, M.; Allan, J.D.; Meyer, J.; Bernhardt, E.S. River restoration in the twenty-first century: Data and experiential knowledge to inform future efforts. *Restor. Ecol.* **2007**, *15*, 472–481. [CrossRef]

114. Stephens, T.; Brown, D.; Thornley, N. *Measuring Conservation Achievement: Concepts and Their Application over the Twizel area*; Department of Conservation: Wellington, New Zealand, 2002.

115. Mansfield, B.; Towns, D.R. Lesson of the islands. *Ecol. Restor.* **1997**, *15*, 138–146.

116. Lee, M.; Hancock, P. Restoration and stewardship volunteerism. In *Human Dimensions of Ecological Restoration: Integrating Science, Nature, and Culture*; Egan, D., Hjerpe, E.E., Abrams, J., Eds.; Island Press/Center for Resource Economics: Washington, DC, USA, 2011; pp. 23–38.

117. Arlettaz, R.; Schaub, M.; Fournier, J.; Reichlin, T.S.; Sierro, A.; Watson, J.E.M.; Braunisch, V. From publications to public actions: When conservation biologists bridge the gap between research and implementation. *BioScience* **2010**, *60*, 835–842. [CrossRef]

118. Innes, J.; Burns, B.; Sanders, A.; Hayward, M.W. The impacts of private sanctuary networks on reintroduction programs. In *Advances in Reintroduction Biology of Australian and New Zealand Fauna*; Armstrong, D.P., Hayward, M.W., Moro, D., Seddon, P.J., Eds.; CSIRO Publishing: Melbourne, Australia, 2015; pp. 185–200.

119. Waldron, A.; Mooers, A.O.; Miller, D.C.; Nibbelink, N.; Redding, D.; Kuhn, T.S.; Roberts, J.T.; Gittleman, J.L. Targeting global conservation funding to limit immediate biodiversity declines. *Proc. Natl. Acad. Sci. USA* **2013**, *110*, 12144–12148. [CrossRef] [PubMed]

land

MDPI

Article

Terrestrial Species in Protected Areas and Community-Managed Lands in Arunachal Pradesh, Northeast India

Nandini Velho [1,*], Rachakonda Sreekar [2] and William F. Laurance [1]

[1] Centre for Tropical Environmental and Sustainability Science (TESS) and College of Marine and Environmental Sciences, James Cook University, Cairns, QLD 4878, Australia; bill.laurance@jcu.edu.au

[2] School of Biological Sciences, University of Adelaide, Adelaide, SA 5000, Australia; rachakonda.sreekar@outlook.com

* Correspondence: nandinivelho@gmail.com

Academic Editors: Jeffrey Sayer and Chris Margules
Received: 19 August 2016; Accepted: 20 October 2016; Published: 26 October 2016

Abstract: Protected areas (including areas that are nominally fully protected and those managed for multiple uses) encompass about a quarter of the total tropical forest estate. Despite growing interest in the relative value of community-managed lands and protected areas, knowledge about the biodiversity value that each sustains remains scarce in the biodiversity-rich tropics. We investigated the species occurrence of a suite of mammal and pheasant species across four protected areas and nearby community-managed lands in a biodiversity hotspot in northeast India. Over 2.5 years we walked 98 transects (half of which were resampled on a second occasion) across the four paired sites. In addition, we interviewed 84 key informants to understand their perceptions of species trends in these two management regimes. We found that protected areas had higher overall species richness and were important for species that were apparently declining in occurrence. On a site-specific basis, community-managed lands had species richness and occurrences comparable to those of a protected area, and in one case their relative abundances of mammals were higher. Interviewees indicated declines in the abundances of larger-bodied species in community-managed lands. Their observations agreed with our field surveys for certain key, large-bodied species, such as gaur and sambar, which generally occurred less in community-managed lands. Hence, the degree to which protected areas and community-managed lands protect wildlife species depends upon the species in question, with larger-bodied species usually faring better within protected areas.

Keywords: Arunachal Pradesh; community-managed lands; gaur; India; management; parks; patrolling; sambar; tigers

1. Introduction

Terrestrial protected areas cover 15.4% of the world's land area [1]. Within the tropics about a quarter of all forested lands are afforded some degree of protection [2], but the type and extent of protection varies geographically. While land afforded at least some protection is much higher in South and Central America (25%–28% of the land is listed in IUCN Categories I–VI, which includes all types of formal protected areas), the figure is much lower in Asia (12.4%) [1]. Notably, the area of land intended to receive strict protection in the tropics is uniformly low in both the Neotropics and Indo-Malayan region [3]. ; The diversity of land-protection regimes in the tropics provides a setting in which to understand the relative importance of protected areas compared to adjacent forests that are often managed by resident communities. Furthermore, the relatively small fraction of strictly protected areas can often make their ecological representation, species populations, and biodiversity patterns inadequate within the entire protected-area network [4,5]. However, protected areas are important reservoirs for maintaining biodiversity [6] and reducing deforestation [7]. They are often considered the first line of defence for wildlife protection [8] and have important values for biodiversity and community well-being [9].

Continuing anthropogenic pressures arising from habitat loss, fragmentation, and hunting are serious challenges in protected areas [10–12]. As protected areas continue to be degraded and adjacent community-managed lands are converted for agriculture and other human uses, obtaining on-ground information about the relative biodiversity values of protected areas and community-managed lands remains a crucial challenge. Past evaluations have largely focused on forest cover [7,13] and abiotic pressures such as fire frequency [2] as proxies for reserve "health", but often these comparisons are geographically unmatched—limiting one's confidence in their conclusions. However, researchers are recognizing that comparisons should be matched [2,14,15], with a focus on expanding metrics to understand the responses of animal species [14]. At the same time, the evidence supporting the efficacy of community-forest management remains weak because of a paucity of rigorously designed studies [16].

An understanding of protected areas versus community-managed lands is especially important in a highly populous, megadiverse developing country such as India. This nation sustains half of the world's tigers, 60 percent of all Asiatic elephants, and 70 percent of all one-horned rhinoceros [17,18]. Although the threats to biodiversity vary widely by species and region, habitat loss and degradation and hunting remain the predominant stresses to Indian biodiversity [6,19].

Arunachal Pradesh, in northeast India, harbours two global biodiversity hotspots, and has the second-highest level of biodiversity globally, after the northern Andes [20]. The decline in important mammal species such as tigers are impacted by socioeconomic changes and institutional inadequacies [19,21]. Our study seeks to addresses how much biodiversity (defined as a range of detectable mammals and pheasants) is harboured in formally protected areas versus adjacent community-managed lands.

Specifically, we used transect-based animal-sign surveys in conjunction with interviews with local residents to assess the persistence of a range of mammal species in each management regime across four independent sites, using a paired study design. We predicted that (a) occurrence of larger-bodied species would be higher in protected areas and (b) key-informant observations will reflect species occurrences in each of these two management regimes relatively accurately.

2. Study Area

Our study area spanned four independent, paired sites (eight sites in total, Figure 1) in the Kameng Protected Area Complex, which, at 3500 km^2, is the largest contiguous forest tract in the Eastern Himalaya Global Biodiversity Hotspot. Historically, there has been no formal land-tenure system in the state, apart from the established hierarchy of ownership rights among tribes: individual, family, clan, village, and tribe. Different tribes and clans have clearly demarcated land management and inter-community land boundaries that operate as areas of management. The relatively recent Arunachal Pradesh (Land Settlement and Records) Act, 2000 and Scheduled Tribes and Other Traditional Forest Dwellers (Recognition of Forest Rights) Act, 2006 tried to formalise land-tenure for individuals, but customary rights are still exercised by different tribes. In the past, although parcels of land may have been privately owned in community-managed lands, there was no formalized system of issuing certificates; ownership was and is still based on an understanding between individuals of the same tribe. However, issuing Land Possession Certificates for private ownership requires inspection and clearance from multiple levels, including the Forest Department, Village Council heads, and the District Administration. Therefore, in the same community-managed lands, there may be unofficial land tenure systems that are based on established mutual and inter-personal agreements and official systems that are based on government records. In this region, protected areas are owned and managed by the state government (they are supposed to restrict encroachment, logging, hunting, and other anthropogenic activities, as mandated by the Wild Life (Protection) Act, 1972), whereas in community forests (or Unclassified State Forests) tribal people exercise their customary rights, that include collecting fuel wood and non-timber forest products. Thus community-managed lands also had more variable land-use patterns. Furthermore, while the Wild Life (Protection) Act, 1972 prohibits hunting of listed species, often cultural and village-level restrictions on hunting vary in the degree of overlap with this legislation.

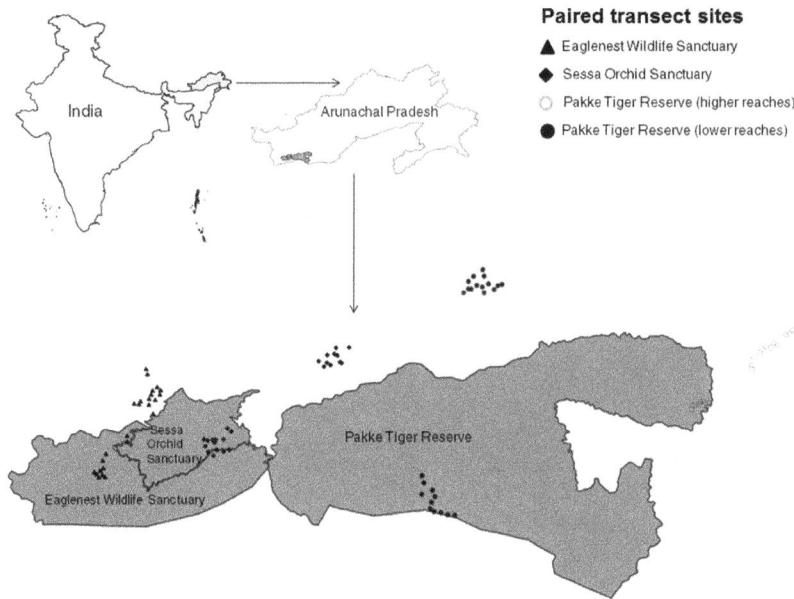

Figure 1. Map of the study area, showing transect locations in the four protected areas (grey) and community-managed lands (white).

We sampled community-managed lands belonging to four tribes, Nyishi, Aka, Bugun, and Shertukpen, between August 2011 and April 2014. Our study spans three protected areas and the adjacent community-managed lands of these four tribes. Thus, we have four independent comparisons of respective community-managed lands: at the lower (median: 667 m a.s.l.) and higher (median: 1346 m a.s.l.) reaches of Pakke Tiger Reserve, at Sessa Orchid Sanctuary, and at Eaglenest Wildlife Sanctuary.

Topographically, the paired sites are similar. The aspect (median: 210.4, 95% CI: 24.6–351), elevation (1155, 95% CI: 142–2785.4) and slope (19.7, 95% CI: 1.6–36) in community-managed lands is similar to that in the protected areas (aspect: 186.8, 95% CI: 18–335.2, elevation: 1417.5, 95% CI: 141–2481.6 and slope: 23.2, 95% CI: 0.7–44.7).

3. Data Collection

3.1. Sign Surveys

Each of these four sites has a road or a path that we used as a sampling backbone. Within each site, we used several 1 km long segments of the path or road as independent units to sample different habitat types: agriculture and fallow; secondary and logged forests; and primary forests (which we define as relatively undisturbed forests with no ongoing anthropogenic modification). For each 1-km segment of road, we walked 500 m long U-shaped transects within the habitat of interest. At each transect, we recorded direct and indirect signs of detectable terrestrial mammals and larger birds, such as pheasants, every 20 m. Two observers walked along each segment a few metres apart, recording these signs independently. In total, we had 44 transects in protected areas and 54 transects in community-managed lands; the sample size for the latter was slightly larger given that they had more varied land-use types than did the protected areas. From October 2011 to March 2013, we walked a total of 98 transects, of which we re-sampled roughly half of these transects (48) at two sites (Eaglenest Wildlife Sanctuary and the lower reaches of Pakke Tiger Reserve).

3.2. Interviews

Given that we were interested in understanding local views towards species abundance trends, we asked respondents to classify species on a scale from −2 (for extirpations) to +2 (for large increases), with a score of zero indicating no change in abundance. We were also interested in comparing the perceptions of perceived abundance trends over time in a community-managed land and protected area. For this comparison, the key informant profiles in the community-managed lands were the same as those mentioned above and mainly belonged to the Aka tribe, with a lower number from the Nyishi tribe. The lower reaches of Pakke Tiger Reserve are uninhabited and so we chose to interview long-term research assistants, people involved in nature-based tourism, forest watchers, and guards. As such, their profiles differ from the typical interviewees at other sites. We acknowledge that population declines might be underestimated within Pakke (because many of the interviewees worked for the Forest Department and might not be comfortable reporting declines), but we hope to have minimised this bias as one of the authors (Nandini Velho) has worked with and become well known to the key informants at Pakke over the past eight years. In this way, we conducted a total of 84 key informant interviews (46 within the protected area and 38 from the community-managed lands) from our paired site in and around Pakke Tiger Reserve.

4. Statistical Analysis

We used Program R (R Development Core Team 2015) for all analyses. In all areas, except within the Eaglenest Wildlife Sanctuary (where domestic hunting dogs and the semi-domesticated cattle-like mythun or mithun *Bos frontalis* were not present), we excluded certain animals from the species-level analysis because it is difficult to distinguish their signs. Specifically, signs of domestic dogs and wild dogs are very similar, as are the signs of the mythun and its wild counterpart, the gaur (*Bos gaurus*). Therefore, these species were excluded from the analyses.

For our analysis of species richness for selected mammals and birds, we used the average number of signs per transect (averaged across two repeats, where applicable) in a transect-by-species matrix. We used the package *vegan* to calculate species richness and perform site-wise comparisons. We report bootstrapped values of species richness and standard errors to make site-wise comparisons among the four sites. The estimated species richness was similar whether we used only fresh signs or all signs, and hence we used the latter for our analysis. Additionally, we constructed rank-abundance plots and fitted log-normal models of species abundance (for community-managed lands and protected areas) for each of the four sites using the "*radfit*" function in the package *vegan* [22]. Higher evenness in the community is represented by shallower estimates (close to zero) of the species abundance distribution. To determine whether protected areas had greater evenness (shallower estimates) than community-managed lands, we followed Gelman & Hill [23]. We used the "*sim*" function in the *arm* package to construct 95% confidence intervals of our parameter estimates by resampling 1000 times from its posterior distribution. An approximate one-tailed *p*-value of two samples was estimated as follows: $p = 1 - (x/n)$, here "x" is the number of samples where estimates of the protected area was greater than the 50th quantile (median) estimate of the community-managed land, and "n" is the total number of resamples [24].

To investigate the relationship between species occurrence and body mass in protected areas and community-managed lands, we used our transect data to get the proportion of segments where each species was encountered. The body mass data for most species was drawn from [25]. We used generalised linear models (GLMs) with binomial error structure to model the effects of body mass on occurrence probability for each site. We used McFadden's pseudo-R^2 to calculate the deviance explained by the GLMs. We then used the method recommended by [23] (see above for details) to determine if the model estimates of protected areas are greater than community-managed lands.

Our interview analysis only included species for which there were at least 10 responses by key informants. If our interviewee was unsure about species identity (or thought it was not found in the area), the species was excluded as a data point from that interview. Using this approach, we had enough information to model population trends of 23 species from within Pakke Tiger Reserve and 29 species from within the adjacent community-managed lands. We used general linear models (GLMs) to determine the effect of body mass on average species scores, as larger animals are often selectively hunted. Again, we used the method recommended by Gelman and Hill [23] to determine if the model estimates for protected forests are greater than for community-managed forests. We used an average body size for similar-looking otter species, two macaque species, and six pheasant species.

5. Results

5.1. Species Richness

In general, protected areas had slightly higher species richness (range of estimated species richness bootstrap: 13–17) than community-managed lands (10–15) (Figure 2). The lower reaches of Pakke Tiger Reserve and Sessa Orchid Sanctuary had higher species richness compared with the community-managed lands, whereas Pakke's higher reaches and Eaglenest had almost similar richness between regimes (Figure 2). Similarly, the species evenness was higher in protected areas than community-managed lands of Pakke's lower reaches ($p < 0.001$) and Sessa ($p < 0.001$), whereas Pakke's higher reaches ($p = 0.46$) and Eaglenest ($p = 1$) had similar evenness (see Figure 3). The community-managed lands of Eaglenest appeared to have higher richness and evenness (Figures 2 and 3). Barking deer, porcupines, small carnivores, and pheasants were frequently encountered on transects in community-managed lands. Elephant, sambar, gaur, and barking deer were frequently encountered in protected area sites, except for Sessa Orchid Sanctuary (Figure 3).

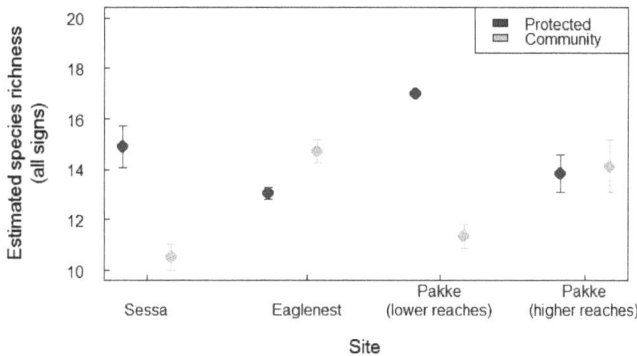

Figure 2. Bootstrapped species richness estimates with standard errors across four independent site comparisons. Pakke Tiger Reserve (lower reaches) had the highest species richness, while the community-managed land adjacent to Sessa had the lowest.

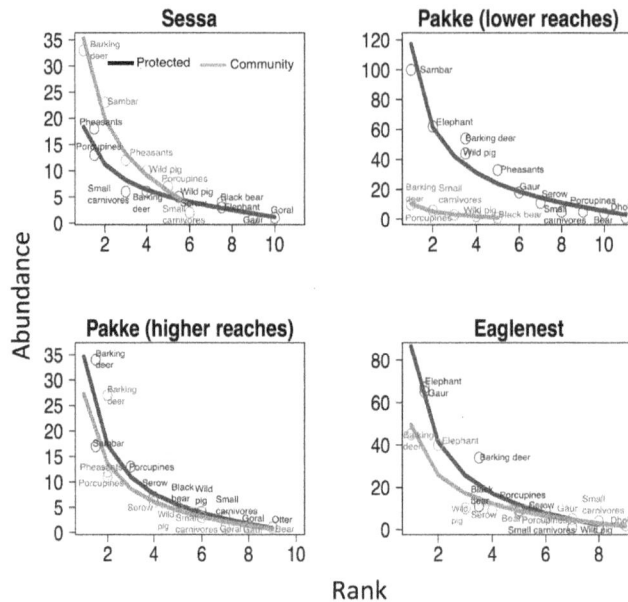

Figure 3. Plots showing species evenness in protected areas and community-managed lands. Species rank abundance distribution plots are fitted with lognormal models. Lines and points are trends and species occurrences in protected areas (dark grey) and community-managed lands (light grey). Protected areas of Sessa and Pakke's lower reaches had higher evenness than their community-managed lands.

5.2. Occurrence

In protected areas, the occurrence probability of terrestrial mammals increased with body mass in the lower reaches of Pakke Tiger Reserve (GLM: $R^2 = 0.05$, $p = 0.018$) and Eaglenest Wildlife Sanctuary ($R^2 = 0.11$, $p < 0.001$). The higher reaches of Pakke Tiger Reserve ($R^2 = 0.04$, $p = 0.124$) did not show any relationship with body mass and Sessa was the only protected area where occurrence of terrestrial mammals decreased with increasing body mass ($R^2 = 0.25$, $p < 0.001$; Figure 4). In contrast, across three of the four community-managed lands the occurrence probability of terrestrial mammals decreased with increasing body mass (Sessa: $R^2 = 0.05$, $p = 0.048$; Pakke lower reaches: $R^2 = 0.20$, $p = 0.002$; Pakke higher reaches: $R^2 = 0.28$, $p < 0.001$). The community-managed lands around Eaglenest were the only exception, where there was no observable relation with body mass ($R^2 = 0.02$, $p = 0.16$; Figure 4). The model estimates were higher in protected areas than community-managed lands in all locations (Pakke LR: $p < 0.001$; Pakke HR: $p = 0.006$; Eaglenest: $p = 0.013$; Figure 4), except for Sessa ($p = 0.92$). This indicates that larger species had steeper declines in community-managed lands at all locations except Sessa.

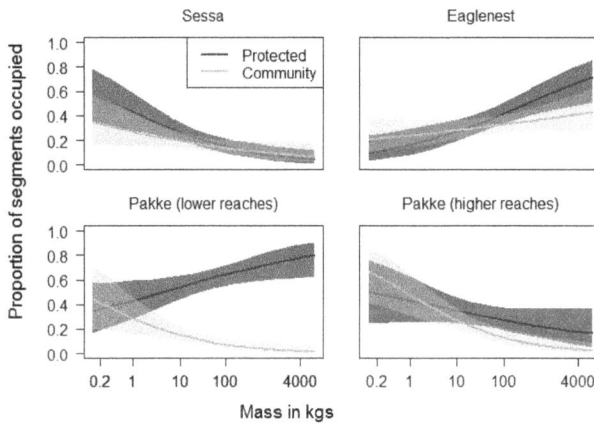

Figure 4. Probability of occurrence of terrestrial vertebrates as a function of body mass. Lines represent model predictions and the shaded dark and light grey regions 95% confidence intervals. The occurrence probability of large-bodied species declined more steeply in community-managed lands than in protected areas, at all locations except Sessa. Note that the x-axis is on a log scale.

5.3. Interviews

When the mean interview scores of species (i.e., the perceived abundance over the last 30 years) were modelled against body size, the estimated ($p < 0.001$) values of Pakke Tiger Reserve were greater than for community-managed lands, indicating a decline in species abundance with body mass in community-managed lands (Figure 5). Interviewees perceived relatively stable abundances of sambar, gaur, elephant, and tiger in the sites around Pakke Tiger Reserve, and associated perceived decreases of these species with adjacent community managed-lands (Figure 5). Qualitatively similar trends were found from our transect data as well (Figure 6).

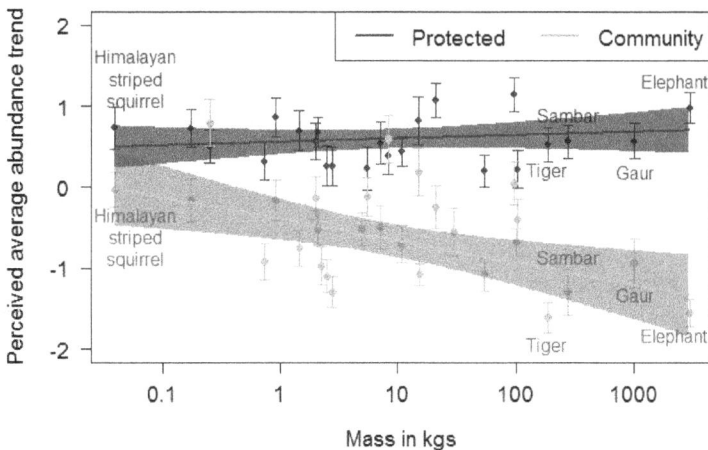

Figure 5. Average scores for perceived species trends based on key informant interviews. The black dots are means and standard errors for each species in Pakke Tiger Reserve, whereas the grey dots are means and standard errors for each species in the adjacent community-managed lands. The solid lines show the fitted ordinary least-squared prediction for species as a function of increasing body mass, and the shaded polygons are the 95% confidence interval. Note that the x-axis is on a log scale. Community-managed lands showed a negative perceived abundance trend with increasing body mass.

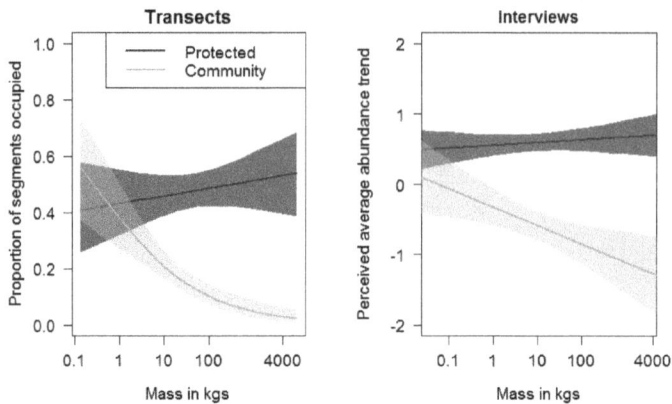

Figure 6. Graphical comparisons of transect and interview data from Pakke Tiger Reserve showing similar patterns of species abundance with body mass. Solid lines and their shaded polygons represent model fits and 95% CIs, respectively. The x-axis is on a log scale.

6. Discussion

Our study shows that protected areas are important for large vertebrates. In general, larger-bodied mammal species had higher occurrences in protected areas than in community forests (Figure 4). We also show that well-managed community lands can be as important as protected areas. For example, vertebrates in community-managed lands and protected areas of Eaglenest and the higher reaches of Pakke had similar richness, evenness, and probability of occurrence (see Figures 2–4).

Key informants report a decline in larger-bodied species in community-managed lands, but not in protected areas (see Figure 5). We acknowledge the limitations of this comparison as the respondents in each area have differing profiles. However, our interviewees' observations of a less steep decline in larger-bodied species in Pakke Tiger Reserve are supported by our field data (Figure 6). Others have also found local knowledge to be a useful tool for biodiversity monitoring [26], though we caution that such data might be difficult to use for elusive and morphologically confusing species. Taken together, our field and interview data indicate that protected areas are important for large-bodied herbivores such as sambar and gaur, and for top predators such as the tiger. These large-bodied species, our interviewees perceived, had declined in abundance in community-managed lands (see Figure 5). Other studies have also noted the relative importance of protected areas for large-bodied species such as elephants [14] and other mammalian and bird species [27,28].

Given that there are multiple reserves within the Kameng Protected Area Complex, future studies should focus on longer-term estimation of key faunal species to identify if and where there are source populations. A positive outcome for people and wildlife may be a mosaic landscape of protected areas in a matrix of community-managed lands. If so, an understanding of dispersal constraints and functional connectivity [29] within the landscape might allow one to better predict the potential of protected areas and community-managed lands to regain some of their extirpated species. Furthermore, there are nuances related to politics, economics, and other social aspects that need to be factored in when considering the strengthening of protected-area networks [30]. This is especially needed as protected areas are considered to be least effective in Asia when compared with other regions [12]. Nonetheless, we find that anti-poaching efforts in the lower reaches of Pakke Tiger Reserve (which also has high anthropogenic pressures) are likely to have benefits for hunted mammals and birds. Faunal species richness and occurrence, moreover, are much lower in surrounding community-managed land than in the reserve itself. At the other end of the spectrum is Sessa Orchid Sanctuary, which has a highway passing through it; here forests are rapidly being converted to other land-uses because of illegal logging and agricultural expansion.

Depending on the context, the matrices of community-managed lands around protected areas hold promise and potential for wildlife conservation. For example, species richness in the higher-altitude community-managed lands of Eaglenest are comparable to the protected area (see Figures 2 and 3). These community-managed lands may have low hunting pressures compared with other sites, most likely because of the prevalence of both Buddhism and nature-based tourism. Although species richness was low in the community-managed land around Sessa Orchid Sanctuary, it still had higher occurrences of species compared with the adjacent protected area, indicating that there might be some pockets outside reserves where species are still found in high abundances. The community-managed lands around Sessa Orchid Sanctuary have low population pressures as many local people have moved to smaller towns that are more centrally located. Thus land-use, land cover, and management activities at each of the paired sites are likely to be an important driver of biodiversity persistence, and merit further investigation.

To our knowledge, our study is one of the first to examine the differences in biodiversity values of protected areas and community-managed lands at multiple paired sites in the tropics. The possibility of finding community-managed lands that harbour significant biodiversity values should be explored further, as such lands and management could complement existing protected areas and may well function better than open-access areas [31,32]. The State Government, through draft legislation (Arunachal Forest Act, 2014), seeks to regulate human extractive use of these community forests (Unclassified State Forests). More importantly, it plans to formulate rules for land-use management, moving to a more centralised governance structure. The reclassification of land should not solely focus on creating more formal and smaller reserves but should also recognise the importance of community-managed lands in the larger landscape as a whole. There are other opportunities to improve governance and land-use planning in many of these community-managed lands so that biodiversity outcomes could be bettered. For example, it would be helpful to incentivize Village Council leaders to work with the Forest Department to help implement regulations related to the hunting of large-bodied species on community-managed lands. In Arunachal Pradesh, 62% of the forests are such community-managed lands [33]. At the same time, active management in protected areas is likely to be beneficial, especially in areas that have high anthropogenic pressures. Even in the absence of active management, protected areas still retain important biodiversity values on a large scale.

Acknowledgments: We thank Dinesh Subbha, Bharat Tamang, Shambu Rai, Chamu Rai, Mangal Rai, Gorey Rana, Elizabeth Soumya, Meghna Agarwala, Aditi Kulkarni, Dumbar Pradhan, Bhayung Marphew, Lobsang Marphew, Koliya Sarmah, Maran Degio, Miyali Sidisow, Gentlhe Yamhe, Madhu Degio, Panneerselvam Karthikeyan, Bikram Aditya Roy, Priya Singh, Soren Goyari, Radhe Nabam, Nana Nabam, Ranjan Mallick and Chandan Ri for their assistance in the field. Our special thanks to Luis Velho, Nima Tsering Monpa and Putul Sarmah for support and advice, and to Ramana Athreya and Neelam Dutta for useful inputs. Arunachal Forest Department provided permits. Millo Tasser supported our work in Eaglenest and Tana Tapi supported our work in Pakke and other places. Our gratitude to Indi Glow and the Singchung Village Council for their support and enthusiasm while working in the community forests around Eaglenest. Immense help was extended by Pema Mosobi, Nana Khrimey, the Tukpen Village Council, and the Degio clan of Bana. Susan Laurance and Umesh Srinivasan helped refine ideas and supported us throughout this study. We thank Jeffrey Sayers, Matthew Linkie, and Marc Hockings for their comments on this manuscript. The Rufford Foundation partially supported Nandini Velho through a small grant. William F. Laurance thanks the Australian Research Council for funding.

Author Contributions: Conceived and designed the study: N.V. and W.F.L. Data collection: N.V. and W.F.L. Analyzed the data: R.S. and N.V. Wrote the paper: N.V., R.S. and W.F.L.

Conflicts of Interest: The authors declare no conflict of interest.

References

1. Juffe-Bignoli, D.; Burgess, N.; Bingham, H.; Belle, E.; de Lima, M.; Deguignet, M.; Bertzky, B.; Milam, A.; Martinez-Lopez, J.; Lewis, E.; et al. *Protected Planet. Report 2014. Tracking Progress towards Global Targets for Protected Areas*; United Nations Environment Programme World Conservation Monitoring Centre: Cambridge, UK, 2014.

2. Nelson, A.; Chomitz, K.M. Effectiveness of strict vs. multiple use protected areas in reducing tropical forest fires: A global analysis using matching methods. *PLoS ONE* **2011**, *6*, e22722. [CrossRef] [PubMed]

3. Jenkins, C.N.; Joppa, L. Expansion of the global terrestrial protected area system. *Biol. Conserv.* **2009**, *142*, 2166–2174. [CrossRef]

4. Rodrigues, A.S.; Andelman, S.J.; Bakarr, M.I.; Boitani, L.; Brooks, T.M.; Cowling, R.M.; Fishpool, L.D.C.; da Fonseca, G.A.B.; Gaston, K.J.; Hoffmann, M.; et al. Effectiveness of the global protected area network in representing species diversity. *Nature* **2004**, *428*, 640–643. [CrossRef] [PubMed]

5. Venter, O.; Fuller, R.A.; Segan, D.B.; Carwardine, J.; Brooks, T.; Butchart, S.H.; Marco, M.D.; Iwamura, T.; Joseph, L.; O'Grady, D.; et al. Targeting global protected area expansion for imperilled biodiversity. *PLoS Biol.* **2014**, *12*, e1001891. [CrossRef] [PubMed]

6. Karanth, K.K.; Nichols, J.D.; Karanth, K.U.; Hines, J.E.; Christensen, N.L. The shrinking ark: Patterns of large mammal extinctions in India. *Proc. R. Soc. Lond. B. Biol. Sci.* **2010**, *277*, 1971–1979. [CrossRef] [PubMed]

7. Nolte, C.; Agrawal, A.; Silvius, K.M.; Soares-Filho, B.S. Governance regime and location influence avoided deforestation success of protected areas in the Brazilian Amazon. *Proc. Natl. Acad. Sci. USA* **2013**, *110*, 4956–4961. [CrossRef] [PubMed]

8. Bruner, A.G.; Gullison, R.E.; Rice, R.E.; Da Fonseca, G.A.B. Effectiveness of parks in protecting tropical biodiversity. *Science* **2001**, *291*, 125–127. [CrossRef] [PubMed]

9. Leverington, F.; Costa, K.L.; Pavese, H.; Lisle, A.; Hockings, M. A global analysis of protected area management effectiveness. *Environ. Manag.* **2010**, *46*, 685–698. [CrossRef] [PubMed]

10. DeFries, R.; Hansen, A.; Newton, A.C.; Hansen, M.C. Increasing isolation of protected areas in tropical forests over the past twenty years. *Ecol. Appl.* **2005**, *15*, 19–26. [CrossRef]

11. Laurance, W.F.; Useche, D.W.; Rendeiro, J.; Kalka, M.; Bradshaw, C.J.A.; Sloan, S.P.; Campbell, M.; Abernethy, K.; Alvarez, P.; Arroyo-Rodriguez, V.; et al. Averting biodiversity collapse in tropical forest protected areas. *Nature* **2012**, *489*, 290–294. [CrossRef] [PubMed]

12. Spracklen, B.D.; Kalamandeen, M.; Galbraith, D.; Gloor, E.; Spracklen, D.V. A global analysis of deforestation in moist tropical forest protected areas. *PLoS ONE* **2015**, *10*, e0143886. [CrossRef] [PubMed]

13. Ellis, E.A.; Porter-Bolland, L. Is community-based forest management more effective than protected areas?: A comparison of land use/land cover change in two neighbouring study areas of the Central Yucatan Peninsula, Mexico. *For. Ecol. Manag.* **2008**, *256*, 1971–1983. [CrossRef]

14. Goswami, V.R.; Sridhara, S.; Medhi, K.; Williams, C.A.; Chellam, R.; Nichols, J.D.; Oli, M.K. Community-managed forests and wildlife-friendly agriculture play a subsidiary but not substitutive role to protected areas for the endangered Asian elephant. *Biol. Conserv.* **2014**, *177*, 74–81. [CrossRef]

15. Carranza, T.; Balmford, A.; Kapos, V.; Manica, A. Protected area effectiveness in reducing conversion in a rapidly vanishing ecosystem: The Brazilian Cerrado. *Conserv. Lett.* **2014**, *7*, 216–223. [CrossRef]

16. Bowler, D.E.; Buyung-Ali, L.M.; Healey, J.R.; Jones, J.P.G.; Knight, T.M.; Pullin, A.S. Does community forest management provide global environmental benefits and improve local welfare? *Front. Ecol. Environ.* **2011**, *10*, 29–36. [CrossRef]

17. Madhusudan, M.D. Living with large wildlife: Livestock and crop depredation by large mammals in the interior villages of Bhadra Tiger Reserve, South India. *Environ. Manag.* **2003**, *31*, 466–475. [CrossRef] [PubMed]

18. Amin, R.; Thomas, K.; Emslie, R.H.; Foose, T.J.; Strien, N.V. An overview of the conservation status of and threats to rhinoceros species in the wild. *Int. Zoo Yearb.* **2006**, *40*, 96–117. [CrossRef]

19. Datta, A.; Anand, M.O.; Naniwadekar, R. Empty forests: Large carnivore and prey abundance in Namdapha National Park, north-east India. *Biol. Conserv.* **2008**, *141*, 1429–1435. [CrossRef]

20. Price, T. Eaglenest Wildlife Sanctuary: Pressures on biodiversity. *Am. Nat.* **2012**, *180*, 535–545. [CrossRef] [PubMed]

21. Aiyadurai, A.; Singh, N.; Milner-Gulland, E.J. Wildlife hunting by indigenous tribes: A case study from Arunachal Pradesh, North-East India. *Oryx* **2010**, *44*, 564–572. [CrossRef]

22. Oksanen, J.; Blanchet, F.G.; Kindt, R.; Legendre, P.; Minchin, P.R.; O'Hara, R.B.; Simpson, G.L.; Solymos, P.; Stevens, M.H.H.; Wagner, H. Vegan: Community Ecology Package. Available online: https://CRAN.R-project.org/package=vegan (accessed on 24 July 2016).

23. Gelman, A.; Hill, J. *Data Analysis Using Regression and Multilevel/Hierarchical Models*; Cambridge University Press: New York, NY, USA, 2007.

24. Sreekar, R.; Huang, G.; Zhao, J.; Pasion, B.O.; Yasuda, M.; Zhang, K.; Peabotuwage, I.; Wang, X.; Quan, R.; Slik, J.W.F.; et al. The use of species-area relationships to partition the effects of hunting and deforestation in a fragmented landscape. *Divers. Distrib.* **2015**, *21*, 441–450. [CrossRef]

25. Menon, V. *A Field Guide to Indian Mammals*; Dorling Kindersley: New Delhi, India, 2003.

26. Anadon, J.; Gimenez, A.; Ballestar, R.; Perez, I. Evaluation of local ecological knowledge as a method for collecting extensive data on animal abundance. *Conserv. Biol.* **2009**, *23*, 617–625. [CrossRef] [PubMed]

27. Reddy, G.V.; Karanth, U.K.; Kumar, N.S.; Krishnaswamy, J.; Karanth, K.K. *Recovering Biodiversity in Indian Forests*; Springer: Singapore, 2016.

28. Velho, N.; Srinivasan, U.; Singh, P.; Laurance, W.F. Large mammal use of protected and community-managed lands in a biodiversity hotspot. *Anim. Conserv.* **2016**, *19*, 199–208. [CrossRef]

29. Vasudev, D.; Fletcher, R.J.; Goswami, V.R.; Krishnadas, M. From dispersal constraints to landscape connectivity: Lessons from species distribution modeling. *Ecography* **2015**, *38*, 967–978. [CrossRef]

30. Symes, W.S.; Rao, M.; Mascia, M.B.; Carrasco, L.R. Why do we lose protected areas? Factors influencing protected area downgrading, downsizing and degazettement in the tropics and subtropics. *Glob. Chang. Biol.* **2016**, *22*, 656–665. [CrossRef] [PubMed]

31. Shahabuddin, G.; Rao, M. Do community-conserved areas effectively conserve biological diversity? Global insights and the Indian context. *Biol. Conserv.* **2010**, *143*, 2926–2936. [CrossRef]

32. Rao, M.; Nagendra, H.; Shahabuddin, G.; Carrasco, L.R. Chapter 10. Integrating community-managed areas into protected area systems: The promise of synergies and the reality of trade-offs. In *Protected Areas: Are They Safeguarding Biodiversity?* Wiley Blackwell: Hoboken, NJ, USA, 2016; pp. 169–192.

33. Menon, S.; Pontius, R.G.; Rose, J.; Khan, M.L.; Bawa, K.S. Identifying conservation-priority areas in the tropics: A land-use change modelling approach. *Conserv. Biol.* **2001**, *15*, 501–512. [CrossRef]

land

MDPI

Review

Will Biodiversity Be Conserved in Locally-Managed Forests?

Jeffrey Sayer [1,*], Chris Margules [1,2] and Agni Klintuni Boedhihartono [1]

1   Center for Tropical Environmental and Sustainability Science, James Cook University, Cairns, QLD 4870,
     Australia; chrismargules@gmail.com (C.M.); agni.boedhihartono@jcu.edu.au (A.K.B.)
2   Research Center for Climate Change, University of Indonesia, Kota Depok, Java Barat 16424, Indonesia
*   Correspondence: jeffrey.sayer@jcu.edu.au

Academic Editor: Andrew Millington
Received: 16 November 2016; Accepted: 9 January 2017; Published: 13 January 2017

Abstract: Recent decades have seen a rapid movement towards decentralising forest rights and tenure to local communities and indigenous groups in both developing and developed nations. Attribution of local and community rights to forests appears to be gathering increasing momentum in many tropical developing countries. Greater local control of forest resources is a response to the failure of government agencies to exercise adequate stewardship over forests and to ensure that the values of all stakeholders are adequately protected. We reviewed evidence of the impact of decentralised forest management on the biodiversity values of forests and conclude that special measures are needed to protect these values. There are trade-offs between shorter-term local needs for forest lands and products and longer-term global needs for biodiversity and other environmental values. We present evidence of local forest management leading to declining forest integrity with negative impacts on both local forest users and the global environment. We advocate greater attention to measures to ensure protection of biodiversity in locally-managed forests.

Keywords: community forest management; decentralised forest management; forest biodiversity conservation; indigenous forest management

---

1. Introduction

Deforestation has been a focus of environmental concern for several decades. Recently the rate of deforestation has slowed and in many countries forests are expanding again. Only a small number of less-developed tropical countries continue to lose forests [1,2]. Forest decline does continue in very important tropical forest countries, such as Brazil and Indonesia. The areas still being lost are often of high biodiversity value. [3]. Areas of natural or primary forests—those with minimal human modification—continue to decrease [4]. We argue that the problem is shifting from loss of forest extent to loss of forest quality and ecological integrity [1]. The move to decentralised management is amplifying risks of significant decline in forest quality, particularly, for loss of forest biodiversity.

Just as we are within sight of solving the problem of deforestation by reversing forest decline other major challenges to forest conservation are emerging. Forests are increasingly fragmented and subject to unsustainable harvesting of numerous products. Invasive species are a major threat to forests worldwide. Climate change is placing stress on forest systems in novel ways. The papers in this Special Issue of "Land" [5] are intended to draw attention to some of the special problems posed by the decentralisation of forest management for biodiversity. We contend that decentralization of management is shifting the balance of concern for forests away from their global public goods values towards more local instrumental values. Local communities are inevitably going to be more concerned about the immediate values that they can derive from forest lands and less concerned with values such as carbon retention, biodiversity conservation, and larger scale hydrological functions [6].

Communities and indigenous peoples are asserting their rights to recover control of forests that they have exploited sustainably often for many generations [7]. Over 50% of protected areas have been established on lands that indigenous peoples have traditionally occupied and used [8]. In many parts of the world local and indigenous peoples are challenging the legitimacy of protected areas which they perceive to be "land grabs" by conservationists [9,10]. There are many situations where local peoples have a strong conservation ethic and are powerful advocates for biodiversity conservation [11], but there is surprisingly little empirical evidence for the extent and effectiveness of decentralized management in achieving biodiversity conservation goals. Local people asserting their rights to self-determination consider that their short-term needs to clear forests for agriculture or to hunt and log in forest areas have greater legitimacy than externally-imposed conservation goals. Advocates of local forest management often claim that local communities will conserve biodiversity more effectively than governments but they rarely present empirical evidence to support these claims. Decentralisation of forest management is advancing rapidly in the absence of real evidence that it is effective in conserving biodiversity. Despite the obvious risk in this course of action many are arguing for even greater allocation of forests to indigenous and local peoples [12,13].

Local self-determination and access to local natural resources is a basic human right but it is always necessary to balance local interests with national, regional or even global interests. The perennial dilemma of conserving the public goods values of forests whilst meeting the immediate needs of local residents has been a central theme in forest stewardship throughout history [14,15]. The tension between local self-determination and the conservation of public goods has been a dominant issue in the development of forest conservation strategies in recent years [16]). The changing thinking on conservation is summarised in the following statement from the activist non-governmental organisation, the Forest People's Programme.

> *In recent years, agreements have been made in several international processes clearly implying that conservation initiatives must respect indigenous peoples' rights. Well-known examples are the Durban Accord and Durban Action Plan (World Parks Congress 2003); the resolutions and recommendations of the World Conservation Congresses of the IUCN, and the Programme of Work on Protected Areas of the CBD and other CBD COP Decisions. This new attitude towards conservation is sometimes called the 'new paradigm on conservation' [17].*

Local communities, indigenous peoples, and the conventional conservation community are in conflict over who should best control forest conservation programmes. The International Union for Conservation of Nature has established the Whakatane Mechanism in an attempt to mediate this debate.

> *The aim of the Whakatane Mechanism is to assess the situation in different protected areas around the world and, where people are negatively affected, to propose solutions and implement them. It also celebrates and supports successful partnerships between peoples and protected areas [8].*

The tension over local control does not only apply to protected areas. Many countries designate most of their forests as "permanent forest estate" where strictly-regulated offtakes of timber and non-timber products is allowed, or where hunting is regulated. Local people are claiming rights to these forests and are particularly vigorous in asserting these rights in situations where state forest agencies are complicit in corporate land grabs. There are numerous examples of industrial forests and estate crops impacting negatively on local livelihoods. Communities are justifiably organising themselves to defend these forests against industrial take over. Communities argue that if they are given secure forest tenure they will be empowered to resist industrial land grabs.

Rural populations, especially in the tropics are increasingly connected to the broader economic landscape. Populations are growing rapidly and naturally aspire to greater material well-being. Rural and forest-dwelling peoples are becoming more assertive in defending their rights to the land [11,12]. Powerful advocacy groups have emerged to defend indigenous and local forest rights.

One of the most effective has been the Rights and Resources Institute (RRI) located in Washington DC (http://rightsandresources.org/). A recent study by RRI in collaboration with the World Resources Institute (WRI) and the Woods Hole Research Center [18]) argues that the expansion of local indigenous land rights would be the most cost-effective way of protecting forests, sequestering carbon, and mitigating climate change. Civil society groups are organising to champion indigenous peoples' forest rights in many countries. In Indonesia the non-governmental organisation Aliansi Masyarakat Adat Nusantara (AMAN http://www.aman.or.id/) has lobbied successfully for political support and legal changes to allow local people to claim ownership of forests that were previously state forest land or protected areas.

Demands for more local control of forests appear to be, at least in part, a reaction to a recent historical period when governments have focused almost exclusively on industrial timber harvesting and conversion of forests to agriculture. In many tropical countries governmental laws and forest agencies have paid little attention to the importance of forests in the livelihoods of local people. Historically there has been an ebbing and flowing of the degree to which control of forests is centralised [19].

Traditional "adat" communities in Indonesia, for example, now claim ownership of a major part of the country's forests (http://www.aman.or.id/). In Indonesia, and many other countries, it is claimed that local control of forests, particularly by indigenous peoples, will provide the best mechanism for conserving all forest values. However many observers have noted that decentralisation of management is difficult to achieve in an equitable manner [19–21]. Governments are reluctant to relinquish control of forests, or they do so whilst leaving in place restrictions that hinder local people in obtaining benefits from the forests [22,23]. The difficulties of ensuring equitable governance of any forests, and especially those under local management, are well documented [24,25]. It takes time to put in place the correct mix of regulations and governance structures to ensure that local forest management is sustainable [26]. The appropriate mix of rules and structures will change over time as the development context of the country evolves. In many instances in recent years the rush to empower local communities and to decentralize forest management has led to the allocation of rights to communities without measures to ensure that these communities are held responsible for sustaining all forest values [6]. Community forests have been badly managed in some countries [27] and there are anecdotal accounts of communities selling their forest rights to oil palm companies in Indonesia and to large scale agricultural enterprises in the Amazon.

Studies of the extent to which local control increases or reduces deforestation have shown that the situation is very variable and dependent on local contexts [28]. The papers in this special issue on "Biodiversity in Community-Managed Forests" explore some of the issues and tensions that occur where local communities are given responsibility for those forests that have global biodiversity values.

Local control could facilitate the use of climate change mitigation funds (REDD+) to reduce deforestation [29]. Progress in implementing REDD+ schemes has been slow and potential impacts of REDD+ on biodiversity have received little attention. Aid agencies and international non-governmental organisations have been promoting community forest management as a means to enable communities to benefit from REDD+ funds. Aid agencies appear to have assumed that communities would conserve global forest values such as biodiversity [21,26]. We contend that there is little evidence to support the assumption that local forest stewardship will automatically ensure the maintenance of the public goods values of forests. There are unresolved assumptions about what defines communities and cultures [30]:

> *"Community-based natural resource management discourses produce images of cultures, communities; and resource management practices that are essentialized, timeless, and homogeneous? In their role as advocates of local resource management regimes, NGOs acting on behalf of local communities may, in part, be constituting the entities whose interests they claim to represent. To what extent might such instances of the "invention of community" have positive or problematic consequences? To what extent, and how, do these representations reflect local concerns, NGO preoccupations, or the interests of transnational conservation, human rights, and environmental*

donors? How have descriptions of local communities, culture, law, and environmental management been creatively shaped to fit larger institutional interests?" [30]

2. What Is the Evidence for Biodiversity Gains from Local Management?

There are surprisingly few studies of the effectiveness of local forest management in conserving biodiversity. Kellert et al. [21] concluded that the record is mixed, and in most of the situations they examined community management had been relatively unsuccessful in either delivering local benefits or conserving biodiversity. Brosius et al. [30] raised the concern that the proliferation of movements and agendas on community, territory, conservation, and the indigenous paradigm means that donor institutions and government agencies need to be more careful in making decisions and giving support to different groups. Many international and local organizations have been involved in the setting up of community-based natural resources management programs (CBNRM) in different parts of the world. The success of these programs has been the subject of controversy. The interpretation of the impacts of CBNRM depends upon the perspective of the observer. Conservation institutions, development organizations, and indigenous people's representatives have differing perspectives on what constitutes success. Studies in Nepal have shown that biodiversity conservation has not received much attention in community managed forests and biodiversity has declined or been altered in these forests [31,32]. There are numerous studies of the effectiveness of local management in delivering economic benefits to local communities, but statements about the broader environmental benefits of local management appear mostly to be aspirational. Local management is largely promoted by development practitioners and non-governmental organisations supporting local and indigenous peoples' interests. This community of practitioners does not have specific expertise or interest in biological diversity and appears often to equate maintenance of tree cover with biodiversity conservation.

Conservation biologists do not appear to have engaged significantly with local management advocates—they choose to work mainly in locations that are uninhabited or are within protected areas. Conservation biologists have the skills and competencies to measure the effectiveness of local management in conserving biodiversity but with a few exceptions they have not addressed this issue [15,21,28]. One notable exception is that conservation biologists have engaged strongly with local communities in areas of known high biodiversity value, and particularly around protected areas. There is a long and mixed history of attempts to promote local development around protected areas in order to reduce pressures on those areas [33]. The recent interest in conservation landscapes provides another example of constructive interactions between conservation biologists and local communities [34]. Conservation landscapes are now included in international lists of protected areas maintained by the International Union for Conservation of Nature, but as the paper by Dudley et al [34]. in this volume shows, there is no systematic effort to assess their biodiversity benefits.

Sustainable wildlife harvesting for food and for sport is well documented and its impacts on biodiversity well known, but controversial. Wild animal populations can sometimes sustain quite heavy levels of harvesting [35], but certain species are much more susceptible than others, and may rapidly decline to extinction [36]. Reductions in populations of larger birds and mammals that have important roles in seed dispersal in forests can lead to rapid declines in tree species diversity and to great reductions in overall forest biodiversity (Terborgh and Perez this volume; [37]). Biodiversity conservation, in many countries, has benefitted from measures taken to improve wildlife habitats for sport hunting, although such measures may not benefit the full range of native biodiversity—predators, for instance, may be eliminated to favour the quarry species sought by hunters.

Initiatives in more developed countries often do benefit biodiversity in locally-managed forest areas. In situations with well-educated and prosperous people, citizen science emerges and people often self-organise to support biodiversity conservation efforts—often in areas under local management [38,39]. Consortia of conservation organisations, governments, and industry can collaborate to conserve biodiversity in areas under local management [40].

The emerging trend towards integrated landscape approaches to balancing conservation and development shows considerable promise for enabling conservation to occur in complex landscapes that include locally-managed areas [41]. However even though landscape approaches are widely used by organisations that have biodiversity conservation as a major part of their mandate there is, as yet, little empirical evidence for their effectiveness in conserving biodiversity [42].

One review of the literature on biodiversity in forests placed under local control [43] concluded that numerous studies showed short-term biodiversity losses. The study concluded that significant areas needed to be closed to any use, but also argued that, in many situations, local management offered better prospects for long-term sustainability of conservation efforts [43].

Several authors have pointed out that the naive perception of community as a homogeneous group having a common vision of conservation is incorrect [20]. Studies have emphasized that *"there is need for a critical perspective of "who actually conserves" in various conservation activities. Community-based conservation programmes should recognise the internal differentiation within the communities"* [44]. Inevitably there is an inequitable distribution of rights and responsibilities for natural resource management in any community. Community-based conservation may be here to stay. The question is how a community's involvement can be made effective. Protection of biodiversity must be based on a wide range of approaches to develop a shared understanding of compatible conservation and development goals at various levels [44].

3. Conclusions

Local management is not a panacea and its success or failure is highly context specific. Numerous studies have highlighted the conditions under which community management tends to be successful [26,45], but they almost all focus on local livelihood benefits and do not explicitly address global public goods values, such as biodiversity. The present expansion of initiatives to devolve forest management to local and indigenous communities clearly requires that more effort is made to assess the impact of local management on biodiversity values. Experience in richer countries suggests that biodiversity values can be maintained in areas under local management when those local managers are motivated to do so or when strict conservation laws are imposed. In poor countries where people are struggling to survive it would be unwise to assume that global environmental values will be a priority. Biodiversity conservation will take second place to immediate development needs in these poorer countries. One could argue that biodiversity should not be a priority in situations where local people are living in poverty. However, the long-term sustainability of local peoples' forests will be in jeopardy if the biodiversity of the forests declines. It will be necessary to have significant areas excluded from human use if biodiversity is to persist. If the allocation of land to these protected areas impacts negatively on local livelihoods, then the global community should compensate local people for the costs they incur. The moral requirement to make such payments for ecosystem services is manifest, but there has been remarkably little progress in putting into place payment schemes for measures to conserve biodiversity [46]. The alternative is to impose strict rules to protect biodiversity in those areas placed under local management. The rush to devolve management is occurring without sufficient attention being given to the need for such special conservation measures. There is a dangerous assumption that local people will protect wild species and will not pose a risk to biodiversity but this has repeatedly been shown to be untrue [3,47] (Terborgh, this volume). Achieving the correct balance between collective rights and individual rights is important when rights to natural resources are allocated. Brosius et al. [30] emphasize that the tension between collective rights and individual rights should receive more attention in moves towards political decentralization and local autonomy. They question "What are the consequences of recognizing community autonomy for larger visions of pluralist civil society? When "natives" become privileged, are other social groups marginalized? What space is there for mobility, migration, and the movements of both rural and urban poor?" [30].

A meta-study of 101 conservation initiatives in territories of indigenous people from 2002 to 2012 assessed whether these conservation initiatives benefitted people and whether collaboration between indigenous groups and outside agencies improved indigenous economies and protected the environment [47]. The study showed that many communities have suffered displacement and increased poverty as a result of losing lands and resources to conservation authorities.

We contend that the move towards greater degrees of local management is, in general, a good thing. However, until such a time as people have met their material needs and have the education and leisure to value biodiversity, they will not conserve the full range of wild species. We propose the following nine measures that would help to ensure that biodiversity goals are met when forest management is decentralised, the measures reflect those that have been advocated for the management of common property resources [48,49] and others that have been identified as preconditions for the success of integrated landscape approaches to conservation and development [39].

- Clear rules defining both rights and responsibilities must be in place. The reality that there will always be a divergence of views amongst different stakeholders must be recognised and addressed.
- Public goods values have to be identified and clarified. The broader environmental objectives for locally-managed areas must be made explicit and measures must be put in place to ensure that these values are maintained.
- A neutral forum for resolving conflicts and reconciling trade-offs between local and public benefits must be established. A process must be in place to enable advocates for local benefits to engage in dialogue with advocates for broader biodiversity and other environmental values.
- Effective compensation mechanisms must be in place to pay local people for the opportunity costs they incur when biodiversity measures conflict with the local use of forests.
- Contributions of locally-managed forests to broader landscape values must be made explicit. Landscape approaches provide a tool for optimising biodiversity benefits of locally-managed areas by understanding the role of these areas in the broader landscape. Locally-managed areas may act as buffers around protected areas or may provide corridors linking natural areas. Locally-managed areas may provide better biodiversity benefits if they are located adjacent to refuge areas [38,41].
- Assessment, monitoring, and adaptive management must be implemented. Local management must provide for assessment of biodiversity values and for monitoring and understanding changes in biodiversity. Measures must be in place to allow for management to be adapted to meet specific needs of biodiversity conservation.
- Legally-mandated institutions must be in place to oversee local management and to ensure that the public goods values of locally-managed forests are protected.
- Special attention must be given to the interests of people practicing traditional lifestyles and belief systems as their needs and potential will differ from those of people who are already part of the cash economy.

The move to local management and the emergence of the "new paradigm for conservation" can provide a basis for better conservation of forest resources including biodiversity [40,50]). However, as the studies in this volume show, the present movement to hand over forest management to communities when checks and balances are not in place to ensure protection of public goods values, such as biodiversity, is dangerous (Terborgh and Peres, this volume; Langston et al. [51]) and runs the risk of serious depletion of the world's biodiversity, especially the numerous unique species of tropical forests. The loss of some of these species may lead to a decline in ecosystem functions and a progressive degradation of the forest and its broader values, including the values that local people obtain from the forests. The principles outlined above must be rapidly applied in tropical developing countries that are at the forefront of the move to decentralised management.

Acknowledgments: We thank the numerous local communities with whom we have worked and whose experience with local forest management has inspired this study. Staff of the Samdhana Institute and Birdlife Indonesia, World Wildlife fund Indonesia and in the Congo Basin and Conservation International in Indonesia for discussions and access to field sites that enriched our understanding of decentralised forest management. We thank Chip Fay, Marcus Colchester, Jatna Supriatna, Nonette Royo, Dominique Endamana and Leonard Usongo for valuable discussions of community and indigenous management.

Author Contributions: The authors reviewed the literature and contributed their personal experience of biodiversity outcomes in forests managed by local communities in tropical countries. Sayer drafted the paper. Margules commented on this draft and added some ideas.

Conflicts of Interest: The authors declare no conflict of interest.

References

1. Sloan, S.; Sayer, J.A. Forest resources assessment of 2015 shows positive global trends but forest loss and degradation persist in poor tropical countries. *For. Ecol. Manag.* **2015**, *352*, 134–145. [CrossRef]

2. Keenan, R.J.; Reams, G.A.; Achard, F.; de Freitas, J.V.; Grainger, A.; Lindquist, E. Dynamics of global forest area: Results from the FAO global forest resources assessment 2015. *For. Ecol. Manag.* **2015**, *352*, 9–20. [CrossRef]

3. Supriatna, J.; Mariati, S. Degradation of primate habitat at Tesso Nilo Forest with special emphasis on the Riau pale-thighed surili (*Presbytis siamensis cana*). *J. Environ. Prot.* **2014**, *5*, 1145–1152. [CrossRef]

4. Morales-Hidalgo, D.; Oswalt, S.N.; Somanathan, E. Status and trends in global primary forest, protected areas, and areas designated for conservation of biodiversity from the global forest resources assessment 2015. *For. Ecol. Manag.* **2015**, *352*, 68–77. [CrossRef]

5. Biodiversity in Locally Managed Lands. Available online: http://www.mdpi.com/journal/land/special_issues/biodiversity_managed (accessed on 11 January 2017).

6. Sayer, J. *The Peoples' Forest Balancing Local and Global Values*; Universidad Autónoma de Madrid: Madrid, Spain, 2007.

7. Agrawal, A.; Chhatre, A.; Hardin, R. Changing governance of the world's forests. *Science* **2008**, *320*, 1460–1462. [CrossRef] [PubMed]

8. Whakatane Mechanism. Available online: http://whakatane-mechanism.org (accessed on 26 October 2016).

9. Cernea, M.M.; Schmidt-Soltau, K. The end of forcible displacements? Conservation must not impoverish people. Section I: The complexities of governing protected areas. In *IUCN Commission on Environmental, Economic and Social Policy*; IUCN: Gland, Switzerland, 2003; pp. 6–101.

10. Cernea, M.M.; Schmidt-Soltau, K. Poverty risks and national parks: Policy issues in conservation and resettlement. *World Dev.* **2006**, *34*, 1808–1830. [CrossRef]

11. Sheil, D.; Puri, R.; Wan, M.; Basuki, I.; Heist, M.V.; Liswanti, N.; Rachmatika, I.; Samsoedin, I. Recognizing local people's priorities for tropical forest biodiversity. *J. Hum. Environ.* **2006**, *35*, 17–24. [CrossRef]

12. Colchester, M. *Forest Peoples, Customary Use and State Forests: The Case for Reform*; Forest People's Programme: Oxford, UK, 2009.

13. White, A.; Martin, A. *Who Owns the World's Forests*; Forest Trends: Washington, DC, USA, 2002.

14. Harrison, R.P. *Forests: The Shadow of Civilization*; University of Chicago Press: Chicago, IL, USA, 2009.

15. Scott, J.C. *Seeing Like a State: How Certain Schemes to Improve the Human Condition Have Failed*; Yale University Press: London, UK, 1998.

16. Sayer, J.; Elliott, C.; Barrow, E.; Gretzinger, S.; Maginnis, S.; McShane, T.; Shepherd, G.; Colfer, C.; Capistrano, D. Implications for biodiversity conservation of decentralized forest resources management. In *Politics of Decentralization, Forests, People and Power*; Earthscan Publications: London, UK, 2005; pp. 121–137.

17. Forest People Praagramme. Available online: http://www.forestpeoples.org (accessed on 28 October 2016).

18. Frechette, A.; Reytar, K.; Saini, S.; Walker, W. *Toward a Global Baseline of Carbon Storage in Collective Lands: An Updated Analysis of Indigenous Peoples' and Local Communities Contributions to Climate Change Mitigation*; Rights & Resources Institute: Washington, DC, USA, 2016.

19. Ribot, J.C.; Agrawal, A.; Larson, A.M. Recentralizing while decentralizing: How national governments reappropriate forest resources. *World Dev.* **2006**, *34*, 1864–1886. [CrossRef]

20. Agrawal, A.; Gibson, C.C. Enchantment and disenchantment: The role of community in natural resource conservation. *World Dev.* **1999**, *27*, 629–649. [CrossRef]

21. Kellert, S.R.; Mehta, J.N.; Ebbin, S.A.; Lichtenfeld, L.L. Community natural resource management: Promise, rhetoric, and reality. *Soc. Nat. Resour.* **2000**, *13*, 705–715.

22. Sarin, M. Joint forest management in India: Achievements and unaddressed challenges. *Unasylva* **1995**, *46*, 30–36.

23. Sarin, M.; Ray, L. *Who Is Gaining? Who Is Losing? Gender and Equity Concerns in Joint Forest Management*; Society for Promotion of Wastelands Development: New Delhi, India, 1998.

24. Larson, A.M.; Barry, D.; Dahal, G.R. New rights for forest-based communities? Understanding processes of forest tenure reform. *Int. For. Rev.* **2010**, *12*, 78–96. [CrossRef]

25. Otto, J.; Zerner, C.; Robinson, J.; Donovan, R.; Lavelle, M.; Villarreal, R.; Salafsky, N.; Alcorn, J.; Seymour, F.; Kleyneyer, C. *Natural Connections: Perspectives in Community-Based Conservation*; Island Press: Washington, DC, USA, 2013.

26. Gilmour, D. *Forty Years of Community-Based Forestry: A Review of Its Extent and Effectiveness*; FAO: Rome, Italy, 2016.

27. De Blas, D.E.; Ruiz-Pérez, M.; Vermeulen, C. Management conflicts in cameroonian community forests. *Ecol. Soc.* **2011**, *16*, 8. [CrossRef]

28. Robinson, B.E.; Holland, M.B.; Naughton-Treves, L. Does secure land tenure save forests? A meta-analysis of the relationship between land tenure and tropical deforestation. *Glob. Environ. Chang.* **2014**, *29*, 281–293. [CrossRef]

29. Larson, A.M. Forest tenure reform in the age of climate change: Lessons for REDD+. *Glob. Environ. Chang.* **2011**, *21*, 540–549. [CrossRef]

30. Brosius, J.P.; Tsing, A.L.; Zerner, C. Representing communities: Histories and politics of community-based natural resource management. *Soc. Nat. Resour.* **1998**, *11*, 157–168. [CrossRef]

31. Acharya, K.P. Does community forests management supports biodiversity conservation? Evidences from two community forests from the mid hills of Nepal. *J. For. Livelihood* **2004**, *4*, 44–54.

32. Shrestha, U.B.; Shrestha, B.B.; Shrestha, S. Biodiversity conservation in community forests of Nepal: Rhetoric and reality. *Int. J. Biodivers. Conserv.* **2010**, *2*, 98–104.

33. Sayer, J.; Campbell, B. *The Science of Sustainable Development: Local Livelihoods and the Global Environment*; Cambridge University Press: Cambridge, UK, 2005.

34. Dudley, N.; Phillips, A.; Amend, T.; Brown, J.; Stolton, S. Evidence for biodiversity conservation in protected landscapes. *Land* **2016**, *5*, 38. [CrossRef]

35. Van Vliet, N.; Milner-Guilland, E.; Bousquet, F.; Saqalli, M.; Nasi, R. Effect of small-scale heterogeneity of prey and hunter distributions on the sustainability of bushmeat hunting. *Conserv. Biol.* **2010**, *24*, 1327–1337. [CrossRef] [PubMed]

36. Terborgh, J. *Requiem for Nature*; Island Press: Washington, DC, USA, 2004.

37. Terborgh, J.; Nuñez-Iturri, G.; Pitman, N.C.; Valverde, F.H.C.; Alvarez, P.; Swamy, V.; Pringle, E.G.; Paine, C. Tree recruitment in an empty forest. *Ecology* **2008**, *89*, 1757–1768. [CrossRef] [PubMed]

38. Sayer, J.; Margules, C.; Bohnet, I.; Boedhihartono, A.; Pierce, R.; Dale, A.; Andrews, K. The role of citizen science in landscape and seascape approaches to integrating conservation and development. *Land* **2015**, *4*, 1200–1212. [CrossRef]

39. Galbraith, M.; Bollard-Breen, B.; Towns, D. The community-conservation conundrum: Is citizen science the answer? *Land* **2016**, *5*, 37. [CrossRef]

40. Hodge, I.; Adams, W. Short-term projects versus adaptive governance: Conflicting demands in the management of ecological restoration. *Land* **2016**. [CrossRef]

41. Sayer, J.; Sunderland, T.; Ghazoul, J.; Pfund, J.-L.; Sheil, D.; Meijaard, E.; Venter, M.; Boedhihartono, A.K.; Day, M.; Garcia, C. Ten principles for a landscape approach to reconciling agriculture, conservation, and other competing land uses. *Proc. Natl. Acad. Sci. USA* **2013**, *110*, 8349–8356. [CrossRef] [PubMed]

42. Sayer, J.A.; Margules, C.; Boedhihartono, A.K.; Sunderland, T.; Langston, J.D.; Reed, J.; Riggs, R.; Buck, L.E.; Campbell, B.M.; Kusters, K. Measuring the effectiveness of landscape approaches to conservation and development. *Sustain. Sci.* **2016**. [CrossRef]

43. Wilshusen, P.R.; Brechin, S.R.; Fortwangler, C.L.; West, P.C. Reinventing a square wheel: Critique of a resurgent "protection paradigm" in international biodiversity conservation. *Soc. Nat. Resour.* **2002**, *15*, 17–40. [CrossRef]

44. Kumar, C. Whither 'community-based' conservation? *Econ. Political Wkly.* **2006**, *41*, 5313–5320.

45. Pagdee, A.; Kim, Y.-S.; Daugherty, P.J. What makes community forest management successful: A meta-study from community forests throughout the world. *Soc. Nat. Resour.* **2006**, *19*, 33–52. [CrossRef]

46. Wunder, S.; Campbell, B.; Frost, P.G.H.; Sayer, J.A.; Iwan, R.; Wollenberg, L. When donors get cold feet: The community conservation concession in setulang (Kalimantan, Indonesia) that never happened. *Ecol. Soc.* **2008**, *13*, 12. [CrossRef]

47. Popova, U. Conservation, traditional knowledge, and indigenous peoples. *Am. Behav. Sci.* **2014**, *58*, 197–214. [CrossRef]

48. Ostrom, E. A general framework for analyzing sustainability of social-ecological systems. *Science* **2009**, *325*, 419–422. [CrossRef] [PubMed]

49. Ostrom, E.; Janssen, M.A.; Anderies, J.M. Going beyond panaceas. *Proc. Natl. Acad. Sci. USA* **2007**, *104*, 15176–15178. [CrossRef] [PubMed]

50. Thackway, R.; Freudenberger, D. Accounting for the drivers that degrade and restore landscape functions in Australia. *Land* **2016**, *5*, 40. [CrossRef]

51. Langston, J.; Riggs, R.; Sururi, Y.; Sunderland, T. Estate crops more attractive than community forests in west Kalimantan, Indonesia. *Land* **2016**, submitted.

Review

Evidence for Biodiversity Conservation in Protected Landscapes

Nigel Dudley [1,2,*], Adrian Phillips [3], Thora Amend [4], Jessica Brown [5] and Sue Stolton [1]

1   Equilibrium Research, 47 The Quays, Spike Island, Cumberland Road, Bristol BS1 6UQ, UK; sue@equilibriumresearch.com
2   School of Geography, Planning and Environmental Management, University of Queensland, Brisbane, QLD 4072, Australia
3   30 Painswick Road, Cheltenham GL50 2HA, UK; adrian.phillips@gmx.com
4   Conservation & Development, Bahnhofstr. 9, Laufenburg 79725, Germany; thora.amend@gmx.net
5   Oldtownhill Associates and IUCN-WCPA Protected Landscape Specialist Group, Newbury, MA 01951, USA; jbrown@oldtownhill.org
*   Correspondence: nigel@equilibriumresearch.com; Tel.: +44-773-454-1913

Academic Editors: Jeffrey Sayer and Chris Margules
Received: 20 September 2016; Accepted: 26 October 2016; Published: 4 November 2016

Abstract: A growing number of protected areas are defined by the International Union for Conservation of Nature (IUCN) as *protected landscapes and seascapes*, or category V protected areas, one of six protected area categories based on management approach. Category V now makes up over half the protected area coverage in Europe, for instance. While the earliest category V areas were designated mainly for their landscape and recreational values, they are increasingly expected also to protect biodiversity. Critics have claimed that they fail to conserve enough biodiversity. The current paper addresses this question by reviewing available evidence for the effectiveness of category V in protecting wild biodiversity by drawing on published information and a set of case studies. Research to date focuses more frequently on changes in vegetation cover than on species, and results are limited and contradictory, suggesting variously that category V protected areas are better than, worse than or the same as more strictly protected categories in terms of conserving biodiversity. This may indicate that differences are not dramatic, or that effectiveness depends on many factors. The need for greater research in this area is highlighted. Research gaps include: (i) comparative studies of conservation success inside and outside category V protected areas; (ii) the contribution that small, strictly protected areas make to the conservation success of surrounding, less strictly protected areas—and vice versa; (iii) the effectiveness of different governance approaches in category V; (iv) a clearer understanding of the impacts of zoning in a protected area; and (v) better understanding of how to implement landscape approaches in and around category V protected areas.

Keywords: protected landscape; IUCN category V; biodiversity conservation

1. Introduction

A growing number of protected areas, particularly—but by no means only in Europe, are defined by the International Union for Conservation of Nature (IUCN) as *protected landscapes/seascapes*, or more formally as *category V protected areas*; "where the interaction of people and nature over time has produced an area of distinct character with significant ecological, biological, cultural and scenic value: and where safeguarding the integrity of this interaction is vital to protecting and sustaining the area and its associated nature conservation and other values" [1]. IUCN identifies six different protected area management categories (see Table 1), of which category V, protected landscapes/seascapes, is the least strictly protected Category VI, the other less strict management regime, gives protection to broadly natural ecosystems which nonetheless provide a sustainable off-take, such as fish or rubber tapped from native trees.

Table 1. The International Union for Conservation of Nature (IUCN) Protected Area Management Categories. Source: Dudley, 2008 [1].

No.	Name	Description
I$_a$	Strict nature reserve	Strictly protected areas set aside to protect biodiversity and also possibly geological/geomorphological features, where human visitation, use and impacts are strictly controlled and limited to ensure protection of the conservation values.
I$_b$	Wilderness area	Usually large unmodified or slightly modified areas, retaining their natural character and influence, without permanent or significant human habitation, which are protected and managed so as to preserve their natural condition.
II	National park	Large natural or near natural areas set aside to protect large-scale ecological processes, along with the species and ecosystems characteristic of the area, which also provide a foundation for environmentally and culturally compatible spiritual, scientific, educational, recreational and visitor opportunities.
III	Natural monument or feature	Areas set aside to protect a specific natural monument, which can be a landform, sea mount, submarine cavern, geological feature such as a cave or even a living feature such as an ancient grove.
IV	Habitat/species management area	Areas that aim to protect particular species or habitats and where management reflects this priority. Many category IV protected areas will need regular, active interventions to address the requirements of particular species or to maintain habitats, but this is not a requirement of the category.
V	Protected landscape or seascape	An area where the interaction of people and nature over time has produced an area of distinct character with significant ecological, biological, cultural and scenic value: and where safeguarding the integrity of this interaction is vital to protecting and sustaining the area and its associated nature conservation and other values.
VI	Protected areas with sustainable use of natural resources	Areas which conserve ecosystems and habitats, together with associated cultural values and traditional natural resource management systems. They are generally large, with most of the area in a natural condition, where a proportion is under sustainable natural resource management and where low-level non-industrial use of natural resources compatible with nature conservation is seen as one of the main aims of the area.

Category V protected areas are neither natural ecosystems nor "wilderness" areas, but rather cultural landscapes that in spite of, or even because of, their long history of human use often contain important biodiversity. Over the past two decades, protected landscapes have increasingly been designated as protected areas in many parts of the world [2]. Currently, there are about 7.3 million km^2 of protected areas reported under IUCN Category V, some 18% of total area of protected areas with an IUCN Category and a considerably higher proportion of terrestrial protected areas [3]. Governments are introducing new designations based on category V management objectives, using a range of governance types [4]. Since 1992 inclusion of a Cultural Landscape designation in the World Heritage Convention has led to the inscription of many World Heritage Cultural Landscapes, with considerable overlap between these sites and category V protected areas [5,6]. In parallel, conservation initiatives such *Satoyama*, a traditional form of land management originating in Japan and now promoted in other countries [7] and designations such as Globally Important Agricultural Heritage Systems are further advancing what might be referred to as a "protected landscape approach" [8], even if they are not all officially recognized as category V protected areas.

The earliest places now recognised as category V protected areas were established mainly to preserve scenic beauty, influenced by the aesthetics of Romanticism [9] and to provide urban populations with access to the countryside. The physical and political battles to access high moors above the cities of Sheffield and Manchester were an important stimulus for the 1949 National Parks and Access to the Countryside Act and helped create the UK's first protected landscape, the Peak

District National Park, designated in 1951 [10]. Throughout Europe [11] and North America [12] the earliest national parks were promoted primarily on the basis of scenery, access and recreation, rather than wildlife conservation [13].

Over time the protected landscape concept has expanded to give greater prominence to nature conservation, cultural heritage, ecosystem services and sustainability models. As countries took a greater interest in biodiversity, managers of protected landscapes have investigated options for strengthening the conservation outcomes. The Satoyama Initiative, launched in Japan in 2010, aims to improve links between biodiversity conservation and cultural landscapes [14]. Natura 2000 legislation in the European Union means that many established protected landscapes are now required to deliver specific conservation objectives that were not always identified when they were first designated.

A seminal workshop held in 1987 in the English Lake District compiled a list of protected landscape values [15], which included the conservation of wild biodiversity; the relative importance of this aim has increased in the years since. A growing number of developing country governments are also attracted by the flexibility that category V offers, rather than more exclusionary conservation approaches, and because they can help to buffer or link more strictly protected areas [16,17]. While "national parks" are often seen in many countries as serving the interests of the rich at the expense of the poor, or as a vestige of colonial occupation, "protected landscapes" can claim to build more on local values and traditions and thus they often receive stronger support. In Madagascar, for instance, a "Malagasy-specific" definition of category V has been developed, suitable for the country's particular cultural and social conditions [18]. As part of Ecuador's new Law of Culture, the designation "Ecuadorian Heritage Cultural Landscape" has been proposed, based on values of Ecuadorian identity and sustaining biocultural diversity and heritage in the Andean sense of "patrimony" [19]. The province of Québec has created a designation called *paysage humanisé* (or "living landscape"), in keeping with category V and modelled after the Regional Nature Parks of France and Belgium, as a means of increasing biodiversity conservation while encouraging sustainable rural development [20].

The IUCN World Commission on Protected Areas has a Specialist Group focused on protected landscapes/seascapes, which serves as a platform for documenting and presenting experience worldwide, mobilizing global expertise, and developing guidance on protected landscapes. It was tasked with providing best practice guidance [21], documenting experience worldwide [8] and investigating different values of category V protected areas such as agrobiodiversity [22] and spiritual values [23].

But the concept has also had important detractors, claiming that protected landscapes are not sufficiently focused on delivering conservation benefits [24]. Critics have claimed that some governments see category V as an "easy option" that does not require major cost, and apply the concept casually and carelessly. In 2005, Locke and Dearden [25] argued that both protected landscapes and sustainable use reserves (IUCN category VI), whilst of cultural and often economic value, had no automatic biodiversity value and should not be "counted" as protected areas. These criticisms generated some considered responses [26], The ensuing debate helped to stimulate a thorough re-examination of the purpose and meaning of protected areas, including protected landscapes, and about the relative conservation benefits of different types of protected areas. Others have argued that the importance of protected landscapes in terms of conserving agro-biodiversity, cultural and spiritual values were additional reasons that justified the approach [22,23]. Critics of the category V protected area model remain sceptical about its value [27].

The biodiversity conservation effectiveness of category V protected areas is particularly important in those places where protected landscapes and seascapes make up a substantial proportion of the conservation estate. For example, in Europe over half the area of protected areas is designated as category V [11], If use of the protected landscape category does not conserve biodiversity then European conservation strategies are in deep trouble [28].

2. Do Protected Landscapes Conserve Biodiversity Effectively?

There are actually two related questions: do category V protected areas have a unique role in protecting culturally adapted wild biodiversity; and do protected landscapes work effectively in protecting wild biodiversity? [29]. Although "biodiversity" is stressed here, implying all ecosystems, species and genetic diversity, it is noted that most studies focus on a few species or particular habitat types.

2.1. Do Category V Protected Areas Have a Unique Role in Conserving Culturally Adapted Wild Biodiversity?

Some researchers regard category V sites as places where wild plant and animal species are so adapted to human management patterns that they will decline if management is removed.

This assumes a unique role for protected landscapes that could not be duplicated in more strictly protected reserves, and remains controversial. For example, many ecologists believe that the millennia-old landscape mosaic of traditional farming in the Mediterranean is now an essential factor in maintaining its biodiversity values. They argue that it is richer in diversity than the original ecosystem [30] and that abandoning (or changing) existing management would reduce biodiversity [31], both of animals [32], and plants [33]. Santos and Thorne [34] identify multi-purpose management as a necessary conservation strategy to avoid "over-maturity" of woodlands and scrub invasion. However, others argue that the emphasis on cultural systems in the Mediterranean under-values ecosystems that develop naturally [35]: a less managed ecosystem will likely have different species but not necessarily of less conservation value. Whether or not cultural landscapes are essential to maintaining biodiversity is therefore partly dependent on the conservation objectives. Category V may have an important role in maintaining those land use systems that give rise to cultural landscapes which are valued for their biodiversity. Examples include: flower-rich chalk pastures grazed by sheep; some upland moorland grazing systems that support good bird habitat; seasonally grazed Alpine pastures with abundant wild flowers; and productive Mediterranean cork oak agro-sylvo-pastoral systems, for example the Iberian dehesa. The role of protected landscapes in sustaining genetic diversity in the form of agrobiodiversity has been documented [36].

Protected landscapes may be important here in providing a policy and economic framework to support these values. Maintaining traditional management to conserve associated biodiversity is only possible if people managing the land agree to adhere to those traditions. Drivers in maintaining traditional (and often economically inefficient) management systems include long-standing cultural expectations, a personal sense of stewardship, and financial or other incentives. Negotiating the various trade-offs involved is a key element of protected landscape management [37].

2.2. Do Protected Landscapes Work Effectively in Protecting Wild Biodiversity?

A more fundamental issue is whether or not a protected landscape approach can help to maintain or restore wild biodiversity in a more general sense. This is particularly important because if a protected landscape is to be recognised as a protected area by IUCN, it must be managed in ways that give priority to the conservation of nature—though "nature" in this context includes geodiversity as well as biodiversity. It is known that protected areas do not invariably protect large mammals [38], but that well-managed protected areas are more effective conservation tools than most other management approaches [39,40]. However, relatively few comparative studies of effectiveness have addressed protected landscapes [41] or compared effectiveness across different IUCN categories.

Category V has to date been relatively under-represented in management effectiveness studies. To compound the problem, most management effectiveness methodologies are weakest in relation to reflecting biodiversity outcomes [42]. Moreover, some of the commonest assessment methods, such as the Management Effectiveness Tracking Tool, METT [43] and RAPPAM [44], base assessment mainly on the opinions of key stakeholders (usually the protected area manager and staff). Many academics, NGOs and governments assume that category V (and category VI) protected areas are less effective in conserving biodiversity than stricter approaches in protected areas [45,46]. NGOs like WWF and Conservation International often simply omit categories V and VI from ecoregional plans and gap analyses, at least in the tropics.

However, a series of meta-studies and individual research projects have suggested that "softer", more community-based approaches can be more effective in conserving biodiversity, at least in some situations, than "harder", exclusionary conservation management.

A study from the World Bank used fire occurrence as a surrogate for deforestation and found that strict protected areas substantially reduced fire incidence in Asia and Latin America, but that multiple use protected areas, including indigenous peoples' reserves, were even more effective [47]. A recent meta-analysis comparing strictly protected areas with community-managed forests (a number of which were also defined as category V protected areas) suggested that the latter had lower and less variable annual deforestation rates [48]. A study across 49 protected areas in 22 countries found protected area category to be insignificant in predicting amount of land clearing [49]. Analysis of 1788 protected areas in Latin America found category V protected areas around the median of all categories in terms of vegetation loss [50]. Joppa et al. [51] looking at natural vegetation cover in protected areas found little difference between different categories, except in West Africa where categories V and VI performed less well than stricter forms of protection. A recent study of species richness and abundance inside and outside protected areas found no significant difference between different groups of categories [52]. The most recent global analysis of protected area performance based on individual management effectiveness studies considered over 8000 assessments of protected areas [53]. Although 86 per cent of protected areas surveyed showed at least some level of effective functioning, this global analysis did not attempt a detailed breakdown of effectiveness by category.

This generally positive link with protected landscapes has not been found in all studies. Andam et al. [54] found deforestation less in category I and II protected areas than in other categories and Bradshaw et al. [46] report that stricter protection is more effective, as did Coetzee et al. [40]. Analysis of threatened species in Australia found that species overlapping category I–IV protected areas had a high number of stable or increasing populations, but found no comparable change in categories V and VI [55]. Comparison of categories in protected areas in four countries—Indonesia, Thailand, Costa Rica and Bolivia—found stricter protection reduced forest loss more than less strict protection, but the differences were not large and sometimes a function of site selection. The authors concluded that strictness of protection is not always more effective [56]. Finally research using the Management Effectiveness Tracking Tool identified a highly significant association between category and management effectiveness, with more strictly protected areas having higher scores for biodiversity conservation—categories I_a, I_b and II being most effective, III and IV in the middle and V and VI least effective [57]. The limitations here are that, as noted above, the METT is relatively weak at measuring biodiversity outcomes; also the proportion of category V protected areas in the sample was very small. METT data therefore need to be treated with some caution here.

Information is even more limited for marine environments. Research on sharks in the Australian Great Barrier Reef Marine Park found that they were only being effectively conserved in the strictly protected (category I_a) zones of the reserve [58]. On the other hand, an analysis of marine protected area (MPA) effectiveness throughout Australia discovered little difference between the categories [59]. A review of over a hundred studies showed strikingly higher fish populations inside the no-take reserves compared with surrounding areas [60], although no-take zones can be, and are, designated within category V MPAs. Effectiveness in MPAs has also been linked to a combination of strong governance structures and community engagement [61] suggesting that management objectives may be less influential than effectiveness of governance.

It should be noted that many studies are limited in the types of biodiversity that they consider, often focusing on forest cover or a few species. Many also tend to concentrate on ecosystems under stress, such as tropical forests, and thus do not provide an unbiased sample of conditions around the world. Furthermore, categories V and VI are assessed together in some studies, although they are markedly different management regimes. All these factors highlight the need for more detailed studies.

Table 2. Examples of protected landscapes and seascapes conserving wild biodiversity. Source: Dudley and Stolton, 2012 [67].

Country	Protected Area	Ecology and Management	Key Species
Croatia	Lonjske Polje Nature Park	Semi-natural floodplains, with pastures, ecotourism connected with storks nesting	Many birds, black stork (*Ciconia nigra*), Eurasian spoonbill (*Platalea leucorodia*).
Spain	Somiedo Natural Park	Mountain pasture, upland agriculture and grazing, ecotourism	Natural forests preventing fragmentation; brown bear (*Ursus arctos arctos*) and capercaillie (*Tetrao urogallus*).
Germany	Lüneburger Heath Nature Park	Heathland area	Black grouse (*Tetrao tetrix*).
Mozambique	Matibane Forest Reserve	Coastal forest area	Threatened forest, especially endemic tree *Icuria dunensis*.
Colombia	Makuira National Park	Conserving forest partly through conservation of sacred places and taboo	Important biogeographical island with high levels of biodiversity, numerous forests including cloud forest.
Mexico	Oaxaca community conservation areas	Wide variety of vegetation types scattered throughout the state	Around 70% of mammal species found in community areas (compared to 60% in strictly protected national parks).
India	Khonoma Nature Conservation and Tragopan Sanctuary	Small area of forest conserved at village edge	Included in an Important Bird Area, species including Blyth's tragopan (*Tragopan blythii*) and mammals such as clouded leopard (*Neofelis nebulosa*).
Canada	Poplar River Initiative	Sustainable hunting reserve of First Nations, aimed for long-term management of beaver and other fur species	Beaver (*Castor canadensis*), lynx (*Lynx canadensis*) and wolf (*Canis lupus*).

There is also a small but growing number of studies of individual category V protected areas. Research in Catalonia, Spain found that protected landscapes provided habitat for rare species, including predators like bear and lynx [62]. Studies by the Royal Society for the Protection of Birds in the UK found that there were quantifiable benefits for wild species in British category V protected areas [63]. Evidence on the positive role of traditional farming methods in conservation has long existed in the Mediterranean region [64]. Research in the Lombardy plain in Italy found that natural habitats declined less, and bird diversity was significantly higher, in protected landscapes than in areas outside protection [65]. The protected landscape approach has more generally been used successfully as the basis for species conservation strategies under the European Union's Natura 2000 network, particularly in the Mediterranean, including maintaining corridors between more strictly protected areas [66]. Dudley and Stolton [67] collected case studies that describe in detail links between protected landscapes and wild biodiversity conservation around the world (Table 2).

3. Discussion

Given the importance attached to protected landscapes and seascapes as conservation vehicles, for example in Europe where they make up over half the protected area estate [11], it is surprising and rather alarming how comparatively few attempts have been made to assess their effectiveness. Furthermore, many of the studies that have taken place are simple comparisons of biodiversity (usually species or ecosystem types) inside and outside protected areas, without taking account of other potential variables. Lack of consideration of counterfactuals further limits confidence in the conclusions. Apart from a few very regionally specific studies, there is still insufficient quantitative evidence about whether or not protected landscapes and seascapes are successful in protecting threatened biodiversity. Even those studies that use the IUCN categories in analysing management effectiveness generally do so by using groups of categories rather than considering individual ones, such as protected landscapes. While the examples collected here show that there are clearly cases where a protected landscape approach has likely been successful (and successful where other stricter approaches would probably have failed), there is also evidence which suggests that this is not always the case. Unless a clearer picture is developed of what is happening in protected landscapes and why, what works and does not work, and the steps that can help to improve the chances of success, there is a risk of seeing further decline in biodiversity.

A series of research projects are needed to fill gaps in knowledge. Amongst the issues that need to be addressed are the following [41]:

(i)   comparative studies of conservation success inside and outside established protected landscapes and seascapes, including comparisons with analogous protected areas in more restrictive management categories;

(ii)   identifying the contribution that small, strictly protected areas or core zones make to the conservation success of surrounding, less strictly protected areas—and vice versa;

(iii) comparison of the effectiveness of different governance types and governance approaches within protected landscapes with respect to both effectiveness of conservation and long-term motivation to protect biodiversity;

(iv)   a clearer understanding of the impacts of zoning within a protected area; and

(v)   a better understanding of how to implement landscape approaches [68]: within protected landscapes and seascapes; between category V protected areas and other protected areas in different categories; and between category V protected areas and surrounding management types which are not protected areas.

(vi)   the specific legal and technical tools, including dedicated monitoring programmes, which are required for protected landscapes management. See Mallarach et al. [26].

This research needs to be conducted in parallel with other actions to strengthen information about protected areas of all kinds. One issue, well recognized but poorly quantified, is that countries apply protected area categories in different ways; some places that are designated as category II are managed more like category V for example, further confusing attempts at understanding relative effectiveness.

While all these analyses will be challenging, none should be impossible. The fact that many protected landscapes are found in the richer countries means that time-series data will often already be present, enabling mapping of the relative success of bird and plant conservation inside and outside protected areas, even at a fine scale. The growing interest in "other effective area-based conservation measures", following their emergence as a topic of investigation from within the Convention on Biological Diversity, means that many governments are adopting more imaginative ways of addressing conservation [69].

In building a research portfolio, category V protected areas offer additional options for working with resident communities in data collection, both in deciding what to measure and how. Recognising that long settled local communities and indigenous peoples often know more about resident wild species than incoming scientists, collaborative approaches that incorporate traditional ecological knowledge will be key. The role of community monitoring in category V protected areas is being developed and refined [70], with work focusing on the best indicators related to particular knowledge within communities [71] and the social process of agreeing indicators [72]. Such monitoring will only be successful if stakeholders are comfortable about sharing information [73], which in turn relates to the overall governance structure, power relations and social interactions within the landscape.

The process of deciding what data to collect can itself be a valuable learning process for managers and communities, as has been recognised in Australia [74].

This review also suggests that there is still much to be learned about management of protected landscapes and that some traditional ideas may need to be modified. "Safeguarding the integrity of the interaction" means more than simply freezing things as they are. It would be fair to say that managing change in protected landscapes remains a challenge that has still to be successfully addressed by many managers and policy makers. In changing conditions, managers might take the cultural landscape as a starting point but then build in deliberate interventions to increase the chances of particular species and groups surviving, whilst acknowledging that management will change over time—as will the biodiversity features. Instead of halting management at a particular historical juncture, innovative category V plans will need to acknowledge that management within a landscape will change and that managers must work, often with many different types of landowners, to implement a landscape approach to conservation: for example altering grazing patterns, retaining old trees, conversely opening up woodland habitats, restoring wetlands, replanting slopes and so on. Protected landscapes play a role here, both within the wider landscape approaches and by providing exemplars of how conservation might take place in other places with longstanding cultural management traditions. The fact that so little is known about the conservation benefits from such sites is a cause for concern. Filling these knowledge gaps needs to be a priority for conservation research over the next few years.

4. Conclusions

The rather limited data available on the effectiveness of protected landscapes and seascapes gives results that suggest variously that they are better than, worse than or the same as more strictly protected categories in terms of conserving wild biodiversity. This may imply that differences are not particularly dramatic, or that effectiveness varies depending on many other factors such as biome, the particular management regime in place, or the attitudes of resident and nearby human communities. It is also clear from this survey that decisive evidence is still lacking.

Category V protected landscapes are often applied in situations where stricter forms of protection are politically or socially impossible; to some extent comparing them with strict reserves is irrelevant because in many cases the choice will be between category V and no protected area at all. Nonetheless, as long as national and regional conservation strategies rely on category V for biodiversity conservation strategies it remains important to know how well they perform at this particular task.

The research gaps identified in the discussion section above, relating to comparative effectiveness of category V, role of smaller reserves, impacts of governance and the most effective management frameworks, all need to be addressed with some urgency. Other issues, such as the potential for protected landscapes to provide connectivity, for instance along migration routes, and the potential for category V areas to enhance socio-ecological resilience in the face of climate and other changes might also usefully be explored. At least some of the information required to build a more complete picture of the role of protected area landscapes and seascapes is probably already available, but—it seems—still needs to be fine-tuned, brought together and analysed. Doing so should help to strengthen management in a type of a protected area that is being used more and more.

Acknowledgments: We are grateful to the German Federal Ministry for Economic Cooperation and Development (BMZ) for funding the set of case studies outlined in Table 2, and to José Alba, David Barton Bray, Fikret Berkes, Miguel Neema Pathak Broome, Brioes-Salas, Elvira Duran, Goran Gugic, Nandita Hazarika, Ben Hoffmann, Jörg Liesen, Rito Mabunda, Agnieszka Pawlowska, Martina Porzelt, Julia Premauer, James Robson, Steve Roeger, Phil Wise, Dijana Zupan and Irina Zupan for preparing the studies. We are also grateful to the members of the World Commission on Protected Areas Protected Landscapes Specialists Group for many discussions on these issues over the last decade. No funding was received for the current paper.

Author Contributions: N.D. wrote the initial text, which was then edited and modified by A.P., T.A., J.B. and S.S.

Conflicts of Interest: The authors declare no conflicts of interest.

References

1.   Dudley, N. *Guidelines for Applying Protected Area Management Categories*; International Union for Conservation of Nature (IUCN): Gland, Switzerland, 2008.
2.   Juffe-Bignoli, D.; Burgess, N.D.; Bingham, H.; Belle, E.M.S.; de Lima, M.G.; Deguignet, M.; Bertzky, B.; Milam, A.N.; Martinez-Lopez, J.; Lewis, E.; et al. *Protected Planet Report 2014*; United Nations Environment Programme-World Conservation Monitoring Centre (UNEP-WCMC): Cambridge, UK, 2014.
3.   UNEP-WCMC; IUCN. *Protected Planet Report 2016*; UNEP-WCMC: Cambridge, UK; IUCN: Gland, Switzerland, 2016.
4.   Brown, J. Bringing together nature and culture: Integrating a landscape approach in protected areas policy and practice. In *Nature Policies and Landscape Policies: Towards an Alliance*; Gambino, R., Peano, A., Eds.; Springer International Publishing: Cham, Germany, 2015; pp. 33–42.
5.   Phillips, A. Cultural landscapes: IUCN'S changing vision of protected areas. In *Cultural Landscapes: The Challenges of Conservation*; Rössler, M., Ed.; World Heritage Centre: Paris, France, 2003; pp. 40–49.
6.   Finke, G. *Landscape Interfaces: World Heritage Cultural Landscapes and IUCN Protected Areas*; International Union for Conservation of Nature (IUCN): Gland, Switzerland, 2013.
7.   Takeuchi, K.; Ichikawa, K.; Elmqvist, T. Satoyama landscape as social-ecological system: Historical changes and future perspective. *Curr. Opin. Environ. Sustain.* **2016**, *19*, 30–39. [CrossRef]
8.   Brown, J.; Mitchell, N.; Beresford, M. *The Protected Landscape Approach: Linking Nature, Culture and Community*; International Union for Conservation of Nature (IUCN): Gland, Switzerland, 2005.
9.   Hourahane, S.; Stolton, S.; Falzon, C.; Dudley, N. Landscape aesthetics and changing cultural values in the British national parks. In *Protected Landscapes and Cultural and Spiritual Values*; Mallarach, J.M., Ed.; Kasparek Verlag: Heidelberg, Germany, 2008; Volume 2, pp. 177–189.
10. Hey, D. Kinder scout and the legend of the mass trespass. *Agric. Hist. Rev.* **2011**, *59*, 199–216.
11. Gambino, R.; Talamo, D.; Thomasset, F. *Parchi d'europa. Verso una Politica Europea per le aree Protette*; ETS Edizioni: Pisa, Italy, 2008.
12. Center, H.F. *The National Parks: Shaping the System*; National Park Service, Department of the Interior: Washington, DC, USA, 2005.
13. Burchardt, J. *Paradise Lost: Rural Idyll and Social Change since 1800*; I.B. Taurus: London, UK; New York, NY, USA, 2002.
14. Kadoya, T.; Washitani, I. The satoyama index: A biodiversity indicator for agricultural landscapes. *Agric. Ecosyst. Environ.* **2011**, *140*, 20–26. [CrossRef]
15. Lucas, P.H.C. *Protected Landscapes: A Guide for Policy Makers and Planners*; Chapman and Hall: London, UK, 1992.
16. Sarmiento, F.O.; Rodríguez, G.; Argumedo, A. Cultural landscapes of the Andes: Indigenous and colono culture, traditional knowledge and ethno-ecological heritage. In *The Protected Landscape Approach: Linking Nature, Culture and Community*; Brown, J., Mitchel, N., Beresford, M., Eds.; International Union for Conservation of Nature (IUCN): Gland, Switzerland; Cambridge, UK, 2005; pp. 143–156.
17. Jones, B.T.B.; Okello, M.; Wishitemi, B.E.L. Pastoralists, conservation and livelihoods in east and southern Africa: Reconciling continuity and change through the protected landscape approach. In *The Protected Landscape Approach: Linking Nature, Culture and Community*; Brown, J., Mitchell, N., Beresford, M., Eds.; International Union for Conservation of Nature (IUCN): Gland, Switzerland, 2005; pp. 107–118.
18. Borrini-Feyerabend, G.; Dudley, N. *Elan durban: Nouvelles Perspectives pour les aires Protégées à Madagascar*; International Union for Conservation of Nature (IUCN); World Commisson on Protected Areas (WCPA); Commission on Environmental, Economic and Social Policy (CEESP): Gland, Switzerland, 2005.
19. Sarmiento, F.; Viteri, X. Discursive heritage: Sustaining Andean cultural landscapes amidst environmental change. In *Conserving Cultural Landscapes: Challenges and New Directions*; Taylor, K., Mitchell, N.J., St. Clair, A., Eds.; Routledge Press: London, UK; Taylor & Francis Group: New York, NY, USA, 2014; pp. 309–324.
20. Blattel, A.; Gagnon, G.; Côté, J.; Brown, J. Conserving agro-biodiversity on the gaspé peninsula of Québec, Canada: A potential role for paysage humanisé designation. In *Protected Landscapes and Agro-Biodiversity Values*; Amend, T., Brown, J., Kothari, A., Phillips, A., Stolton, S., Eds.; Kasparek Verlag: Heidelberg, Germany, 2008; pp. 96–104.

21. Phillips, A. *Management Guidelines for Category V Protected Areas–Protected Landscapes and Seascapes*; International Union for Conservation of Nature (IUCN): Gland, Switzerland; Cambridge, UK, 2002.

22. Amend, T.; Brown, J.; Kothari, A.; Phillips, A.; Stolton, S. *Protected Landscapes and Agrobiodiversity Values*; Kasparek Verlag: Heidelberg, Germany, 2008.

23. Mallarach, J.M. *Protected Landscapes and Cultural and Spiritual Values*; Kasparek Verlag: Heidelberg, Germany, 2008.

24. McEwen, A.; McEwen, M. *National Parks: Conservation or Cosmetics?* George Allen and Unwin: London, UK, 1982.

25. Locke, H.; Dearden, P. Rethinking protected area categories and the new paradigm. *Environ. Conserv.* **2005**, *32*, 1–10. [CrossRef]

26. Mallarach, J.M.; Morrison, J.; Kothari, A.; Sarmiento, F.; Atauri, J.A.; Wishitemi, B. In defense of protected landscapes: A reply to some criticisms of category v protected areas and suggestions for improvement. In *Defining Protected Areas: An International Conference in Almeria, Spain*; Dudley, N., Stolton, S., Eds.; International Union for Conservation of Nature (IUCN): Gland, Switzerland, 2008; pp. 31–37.

27. Shafer, C.L. Cautionary thoughts on IUCN protected area management categories v–vi. *Glob. Ecol. Conserv.* **2015**, *3*, 331–348. [CrossRef]

28. Dudley, N.; Stolton, S. An assessment of the role of protected landscapes in conserving biodiversity in Europe. In *Nature Policies and Landscape Policies: Towards an Alliance*; Gambino, R., Peano, A., Eds.; Springer International Publishing: Cham, Germany, 2015; pp. 315–322.

29. Dudley, N. Why is biodiversity conservation important in protected landscapes? *George Wright Forum* **2009**, *26*, 31–38.

30. Atauri, J.A.; de Lucio, J.V. The role of landscape structure in species richness distribution of birds, amphibians, reptiles and lepidopterans in mediterranean landscapes. *Landsc. Ecol.* **2001**, *16*, 147–159. [CrossRef]

31. González Bernáldez, F. Ecological consequences of the abandonment of traditional land use systems in central Spain. In *Land Abandonment and Its Role in Conservation*; Baudry, J., Bunce, R.G.H., Eds.; Options Méditerranéennes: Série A. Séminaires Méditerranéens; CIHEAM: Zaragoza, Spain, 1991; Volume 15, pp. 23–29.

32. Pino, J.; Rodà, F.; Ribas, J.; Pons, X. Landscape structure and bird species richness: Implications for conservation in rural areas between natural parks. *Landsc. Urban Plan.* **2000**, *49*, 35–48. [CrossRef]

33. Rescia, A.J.; Schmitz, M.F.; Martín de Agar, P.; de Pablo, C.L.; Atauri, J.A.; Pineda, F.D. Influence of landscape complexity and land management on woody plant diversity in Northern Spain. *J. Veg. Sci.* **1994**, *5*, 505–516. [CrossRef]

34. Santos, M.J.; Thorne, J.H. Comparing culture and ecology: Conservation planning of oak woodlands in Mediterranean landscapes of portugal and california. *Environ. Conserv.* **2010**, *37*, 155–168. [CrossRef]

35. Schnitzler, A.; Génot, J.C.; Wintz, M.; Hale, B.W. Naturalness and conservation in France. *J. Agric. Environ. Ethics* **2008**, *21*, 423–436. [CrossRef]

36. Brown, J.; Kothari, A. Traditional agricultural landscapes and community conserved areas: An overview. *Manag. Environ. Qual. Int. J.* **2011**, *22*, 139–153. [CrossRef]

37. Maginnis, S.; Jackson, W.; Dudley, N. Conservation landscapes: Whose landscapes? Whose trade-offs? In *Getting Biodiversity Projects to Work*; McShane, T.O., Wells, M.P., Eds.; Columbia University Press: New York, NY, USA, 2004; pp. 321–339.

38. Craigie, I.D.; Baillie, J.E.M.; Balmford, A.; Carbone, C.; Collen, B.; Green, R.E.; Hutton, J.M. Large mammal population declines in Africa's protected areas. *Biol. Conserv.* **2010**, *143*, 2221–2228. [CrossRef]

39. Geldmann, J.; Barnes, M.; Coad, L.; Craigie, I.D.; Hockings, M.; Burgess, N.D. Effectiveness of terrestrial protected areas in reducing habitat loss and population declines. *Biol. Conserv.* **2013**, *161*, 230–238. [CrossRef]

40. Coetzee, B.W.T.; Gaston, K.J.; Chown, S.L. Local scale comparisons of biodiversity as a test for global protected area ecological performance: A meta-analysis. *PLoS ONE* **2014**, *9*, e105824. [CrossRef] [PubMed]

41. Dudley, N.; Stolton, S. Protected area diversity and potential for improvement. In *Protected Areas: Are They Safeguarding Biodiversity?* Joppa, L., Baillie, J.E.M., Robinson, J.G., Eds.; Wiley Blackwell: Oxford, UK, 2016; pp. 34–48.

42. Hockings, M.; Stolton, S.; Dudley, N.; James, R. Data credibility: What are the "right" data for evaluating management effectiveness of protected areas? *New Dir. Eval.* **2009**, *2009*, 53–63. [CrossRef]

43. Stolton, S.; Hockings, M.; Dudley, N.; MacKinnon, K.; Whitten, T.; Leverington, F. *Management Effectiveness Tracking Tool: Reporting Progress at Protected Area Sites*; World Wide Fund for Nature (WWF): Gland, Switzerland; The World Bank: Washington, DC, USA, 2007.

44. Ervin, J. Rapid assessment of protected area management effectiveness in four countries. *BioScience* **2003**, *53*, 833–841. [CrossRef]

45. Gardner, T.A.; Caro, T.I.M.; Fitzherbert, E.B.; Banda, T.; Lalbhai, P. Conservation value of multiple-use areas in East Africa. *Conserv. Biol.* **2007**, *21*, 1516–1525. [CrossRef] [PubMed]

46. Bradshaw, C.J.A.; Craigie, I.; Laurance, W.F. National emphasis on high-level protection reduces risk of biodiversity decline in tropical forest reserves. *Biol. Conserv.* **2015**, *190*, 115–122. [CrossRef]

47. Nelson, A.; Chomitz, K.M. Effectiveness of strict vs. multiple use protected areas in reducing tropical forest fires: A global analysis using matching methods. *PLoS ONE* **2011**, *6*, e22722. [CrossRef] [PubMed]

48. Porter-Bolland, L.; Ellis, E.A.; Guariguata, M.R.; Ruiz-Mallén, I.; Negrete-Yankelevich, S.; Reyes-García, V. Community managed forests and forest protected areas: An assessment of their conservation effectiveness across the tropics. *For. Ecol. Manag.* **2012**, *268*, 6–17. [CrossRef]

49. Nagendra, H. Do parks work? Impact of protected areas on land cover clearing. *Ambio J. Hum. Environ.* **2008**, *37*, 330–337. [CrossRef]

50. Leisher, C.; Touval, J.; Hess, S.; Boucher, T.; Reymondin, L. Land and forest degradation inside protected areas in Latin America. *Diversity* **2013**, *5*, 779–795. [CrossRef]

51. Joppa, L.N.; Loarie, S.R.; Pimm, S.L. On the protection of "protected areas". *Proc. Natl. Acad. Sci. USA* **2008**, *105*, 6673–6678. [CrossRef] [PubMed]

52. Gray, C.L.; Hill, S.L.L.; Newbold, T.; Hudson, L.N.; Borger, L.; Contu, S.; Hoskins, A.J.; Ferrier, S.; Purvis, A.; Scharlemann, J.P.W. Local biodiversity is higher inside than outside terrestrial protected areas worldwide. *Nat. Commun.* **2016**, *7*, 1–7. [CrossRef] [PubMed]

53. Leverington, F.; Lemos Costa, K.; Courrau, J.; Pavese, H.; Nolte, C.; Marr, M.; Coad, L.; Burgess, N.; Bomhard, B.; Hockings, M. *Management Effectiveness Evaluation in Protected Areas—A Global Study*; The University of Queensland: Brisbane, QLD, Australia, 2010.

54. Andam, K.S.; Ferraro, P.J.; Pfaff, A.; Sanchez-Azofeifa, G.A.; Robalino, J.A. Measuring the effectiveness of protected area networks in reducing deforestation. *Proc. Natl. Acad. Sci. USA* **2008**, *105*, 16089–16094. [CrossRef] [PubMed]

55. Taylor, M.F.J.; Sattler, P.S.; Evans, M.; Fuller, R.A.; Watson, J.E.M.; Possingham, H.P. What works for threatened species recovery? An empirical evaluation for Australia. *Biodivers. Conserv.* **2011**, *20*, 767–777. [CrossRef]

56. Paul, J.F.; Merlin, M.H.; Daniela, A.M.; Gustavo Javier, C.B.; Subhrendu, K.P.; Katharine, R.E.S. More strictly protected areas are not necessarily more protective: Evidence from Bolivia, Costa Rica, Indonesia, and Thailand. *Environ. Res. Lett.* **2013**, *8*, 025011.

57. Dudley, N. *Tracking Progress in Managing Protected Areas around the World: An Analysis of Two Applications of the Management Effectiveness Tracking Tool Developed by WWF and the World Bank*; World Wide Fund For Nature (WWF): Gland, Switzerland, 2007.

58. Robbins, W.D.; Hisano, M.; Connolly, S.R.; Choat, J.H. Ongoing collapse of coral-reef shark populations. *Curr. Biol.* **2006**, *16*, 2314–2319. [CrossRef] [PubMed]

59. Edgar, G.J.; Stuart-Smith, R.D. Ecological effects of marine protected areas on rocky reef communities—A continental-scale analysis. *Mar. Ecol. Prog. Ser.* **2009**, *388*, 51–62. [CrossRef]

60. Halpern, B.S. The impact of marine reserves: Do reserves work and does reserve size matter? *Ecol. Appl.* **2003**, *13*, 117–137. [CrossRef]

61. Pillans, S.; Ortiz, J.C.; Pillans, R.D.; Possingham, H.P. The impact of marine reserves on nekton diversity and community composition in subtropical eastern Australia. *Biol. Conserv.* **2007**, *136*, 455–469. [CrossRef]

62. Mallarach, J.M.; Varga, J.V. *Evaluation of Management Effectiveness of Protected Areas in Catalonia (ei pein deu anys Després: Balanç i Perspectives, Diversitas: 50)*; Universitat de Girona: Girona, Italy, 2004.

63. Robins, M. Protected landscapes: Sleeping giants of English biodiversity. *ECOS* **2008**, *29*, 74–86.

64. Beaufoy, G.; Baldock, D.; Clark, J. *The Nature of Farming: Low Intensity Farming Systems in Nine European Countries*; Institute for European Environmental Policy: London, UK, 1994.

65. Canova, L. Protected areas and landscape conservation in the Lombardy plain (northern Italy): An appraisal. *Landsc. Urban Plan.* **2006**, *74*, 102–109. [CrossRef]

66. De La Guerra, M.M.; Fernández, J.; Alandi, C.; Olmos, P.; Atauri-Mezquida, J.; Montes del Olmo, C. *Territorial Integration of Natural Protected Areas and Ecological Connectivity within Mediterranean Landscapes*; Junta de Andalucía and Red des Espacios Naturales Protegidos de Andalucía: Seville, Spain, 2002.

67. Dudley, N.; Stolton, S. *Protected Landscapes and Wild Biodiversity*; IUCN: Gland, Switzerland, 2012.

68. Sayer, J.; Sunderland, T.; Ghazoul, J.; Pfund, J.L.; Sheil, D.; Meijaard, E.; Venter, M.; Boedhihartono, A.K.; Day, M.; Garcia, C.; et al. Ten principles for a landscape approach to reconciling agriculture, conservation, and other competing land uses. *Proc. Natl. Acad. Sci. USA* **2013**, *110*, 8349–8356. [CrossRef] [PubMed]

69. Jonas, H.D.; Barbuto, V.; Jonas, H.C.; Kothari, A.; Nelson, F. New steps of change: Looking beyond protected areas to consider other effective area-based conservation measures. *Parks* **2014**, *20*, 111–128. [CrossRef]

70. Danielsen, F.; Burgess, N.D.; Jensen, P.M.; Pirhofer-Walzl, K. Environmental monitoring: The scale and speed of implementation varies according to the degree of peoples involvement. *J. Appl. Ecol.* **2010**, *47*, 1166–1168. [CrossRef]

71. Karim, N. Local knowledge of indicator birds: Implications for community-based monitoring in Teknaf Game Reserve. In *Connecting Communities and Conservation: Collaborative Management of Protected Areas in Bangladesh*; Fox, J., Bushley, B.R., Miles, W.B., Quazi, S.A., Eds.; Robbins East-West Centre: Honolulu, HI, USA; Nishorgo Support Project, Bangladesh Forest Department: Dhaka, Bangladesh, 2009; pp. 139–160.

72. Steinmetz, R. *Ecological Surveys, Monitoring and the Involvement of Local Peoples in Protected Areas in Lao PDR*; International Institute for Environment and Development (IIED): London, UK, 2000.

73. Neurauter, J.; Lui, X.; Liao, C. *The role of Traditional Ecological Knowledge in Protected Area Management: A Case Study of Guanyinshan Nature Reserve, Shaanxi, China*; Curtin University of Technology: Perth, WA, Australia, 2009.

74. Izurieta, A.; Sithole, B.; Stacey, N.; Hunter-Xenie, H.; Campbell, B.; Donohoe, P.; Brown, J.; Wilson, L. Developing indicators for monitoring and evaluating joint management effectiveness in protected areas in the northern territory, Australia. *Ecol. Soc.* **2011**, *16*, 9. [CrossRef]

MDPI AG

St. Alban-Anlage 66

4052 Basel, Switzerland

Tel. +41 61 683 77 34

Fax +41 61 302 89 18

http://www.mdpi.com

Land Editorial Office

E-mail: land@mdpi.com

http://www.mdpi.com/journal/land

www.ingramcontent.com/pod-product-compliance
Lightning Source LLC
Chambersburg PA
CBHW041217220326
41597CB00033BA/5996